Corporate
Responses
to
Environmental
Challenges

CORPORATE RESPONSES TO ENVIRONMENTAL CHALLENGES

Initiatives by Multinational Management

ANN RAPPAPORT

and MARGARET FRESHER FLAHERTY

Foreword by William R. Moomaw

Prepared under the auspices of the Center for Environmental Management, Tufts University

Q

QUORUM BOOKS
New York • Westport, Connecticut • London

Library of Congress Cataloging-in-Publication Data

Rappaport, Ann.
   Corporate responses to environmental challenges : initiatives by
multinational management / Ann Rappaport and Margaret Fresher
Flaherty; foreword by William R. Moomaw; prepared under the
auspices of the Center for Environmental Management, Tufts University.
     p.   cm.
  Includes bibliographical references and index.
  ISBN 0-89930-715-9 (alk. paper)
   1. Industry—Environmental aspects—Management—Case studies.
I. Flaherty, Margaret Fresher. II. Tufts University. III. Title.
HD69.P6R36   1992
658.4'08—dc20   91-44706

British Library Cataloguing in Publication Data is available.

Library of Congress Catalog Card Number: 91-44706
ISBN: 0-89930-715-9

First published in 1992

Quorum Books, One Madison Avenue, New York, NY 10010
An imprint of Greenwood Publishing Group, Inc.

Printed in the United States of America

The paper used in this book complies with the
Permanent Paper Standard issued by the National
Information Standards Organization (Z39.48–1984).

10 9 8 7 6 5 4 3 2 1

To Our Parents

Joan Fresher
Robert Fresher
Barbara Rappaport
Raymond Rappaport

Contents

Illustrations

TABLES

FIGURES

Foreword

A series of events has catapulted environmental issues from the local to the global stage. The tragic disaster at Bhopal, India, the Valdez oil spill, the nuclear accident at Chernobyl, and dying forests in Europe and North America demonstrated both the scale of possible environmental damage from "local" events and the potential for such events worldwide. The discovery of the gaping hole in the ozone layer over Antarctica and the greater than anticipated thinning of its protective shield over northern latitudes, the heightened awareness of the likelihood that we are altering the earth's climate by our accelerating release of heat-trapping greenhouse gases, and the documented loss of species and forest lands demonstrated how truly global are the environmental consequences of our everyday actions.

Major events such as these have brought calls for governmental action and international environmental treaties. Yet as important as actions taken by governments acting separately or in cooperation with one another are, it is the individual acts of countless decision makers in the industrial sector that will determine whether or not society is capable of taking effective action. Because of their size and their transboundary nature, multinational corporations (MNCs) are likely to play the most significant role in defining and implementing responses to both global and local environmental concerns. The larger multinationals generate more revenues than all but the major industrial nations, and a single, U.S.-based MNC is responsible for one-quarter of global production of ozone-depleting chlorofluorocarbons (CFCs) and 6 percent of human-induced greenhouse gases.

It is therefore timely to examine the current environmental, health, and safety practices of multinational corporations to gain an understanding of just what kind of role they are currently playing and what kind of leadership we might expect from them in the future. The current study combines survey data that indicates how MNCs view their own environmental performance with systematic, in-depth case studies. The result is an unprecedented insight into the factors that influence environmental practices and decision making within MNCs.

The range of awareness and the degree of concern expressed by representatives of different corporations are surprisingly varied. While some responses are the expected ones, there are many surprises contained in both the survey and the interviews. The fact that (1) less than half of the responding companies cited lower environmental standards abroad as a factor in their decision to locate facilities off-shore, and (2) more than half favored some type of global, environmental standards suggests that at least some MNCs are likely to play a positive role in improving environmental quality. Pressure from the MNC's country of origin clearly plays a role in affecting corporate environmental policies and actions, but as environmental attitudes shift around the world, those firms at the forefront of change may gain a competitive advantage as well. Because of their high profile, MNCs are often under greater scrutiny within a host country to have higher environmental, health, and safety standards than local and state-run enterprises with which they may compete. One possible consequence of this unequal set of expectations is that a subset of MNCs may well become leading agents for improving environmental quality in developing countries, and then press for higher uniform standards. On the other hand, it is clear that a significant minority of the multinationals studied here will continue to lag behind the leaders.

This study represents one of several that has been undertaken by the Center for Environmental Management at Tufts University to gain insights into the environmental practices of corporations and other institutions, and to determine which factors are likely to be most significant in influencing environmentally responsible behavior. This work is to be followed by an analysis of state-owned petroleum companies that will permit a unique comparison with the environmental practices of MNCs identified here.

William R. Moomaw
Director of Research and Policy Development
Center for Environmental Management
Tufts University

Acknowledgments

Many people have contributed to the evolution of this work, however, we alone bear responsibility for shortcomings in the way problems are framed or for any errors in the analysis and interpretation of data.

This book is derived from two related research projects conducted at the Center for Environmental Management, Tufts University. The case studies reported in this book, and the research on which they are based, were a team effort resulting in a report published by the Center, "Global Corporate Environment, Health, and Safety Programs: Management Principles and Practices. " The members of the project team included J. Gary Taylor, Ph.D., Principal, Scharlin/Taylor Associates; Geoffrey Pomeroy; and ourselves. Gary played a critical role in the conceptual development of the research, and much of the wisdom in these pages resulted from his contributions.

We are indebted to the Center for Environmental Management for its support of our work. It was through CEM's Corporate Affiliates Program that both studies were funded. CEM leaders Anthony Cortese and William Moomaw have been strong advocates and constructive critics of our work, and we are grateful for their contributions.

This research is about companies, without whose participation the study could not have occurred. We are indebted to the individuals who represented the case companies and made time for extensive interviews and rigorous inquiry. Our appreciation to them equals our recognition of their courage and willingness to let us examine some of the underpinnings of their companies, operations and philosophies. This was a quantum leap for many corporate cultures, and we are extremely grateful for the time commitments and support given to this project.

We have been assisted immeasurably by a Steering Committee whose interest and support have been unwavering. The committee provided direction and clarity throughout the process of case study project formulation, research design, reviews and editing. Jonathan Plaut, our steering committee chairperson and director of environmental compliance, Allied-Signal Corporation, provided coaching and consistently valuable input. Richard MacLean, manager, environmental protection, corporate environmental programs at General Electric, had an abundance of wisdom as well as a willingness and enthusiasm to share it with us. He provided us with a critical link between academic and corporate audiences.

In addition to Jonathan Plaut and Richard MacLean, other members of the Case Study Project Steering Committee included George Dominguez, Director of Information, Synthetic Organic Chemical Manufacturers Association; Albert Fry, Director, International Environment Bureau; Anthony Cortese, Dean, Environmental Programs, Tufts University; Robert Hollister, Director, Lincoln Filene Center, Tufts University; Arpad von Lazar, Professor of International Energy and Development, Fletcher School of Law and Diplomacy; Denis Simon, Professor of International Business Relations, Fletcher School of Law and Diplomacy; and Rashid Shaikh, Assistant Executive Director, Health Effects Institute, Cambridge, Massachusetts.

Project Advisors were Frank J. Penna, Frank J. Penna Associates, Alvin A. Natkin, President, Continental Environment Company, Inc.

The survey described in this book was also a team effort resulting in a Center for Environmental Management publication, "Multinational Corporations and the Environment: A Survey of Global Practices." The team's principal members were Maureen Hart and ourselves, however, we gratefully acknowledge the enthusiastic and helpful contributions of our colleagues. Assistance in survey design and management of the data, as well as overall encouragement and morale boosting, were provided by Marian Pagano and her associates at the Office of Institutional Research at Tufts University. Much appreciation goes to Geoffrey Pomeroy for his hard work tracking down mailing lists and assisting with the survey design.

We are grateful to Helge Hveem and Audun Ruud of the University of Oslo, Nay Htun of the United Nations Conference on Environment and Development secretariat, and Harris Gleckman of the United Nations Centre on Transnational Corporations for their support and encouragement in the case study research. Dick Mannion from the Foxboro Company, Gary Taylor of Scharlin/Taylor Associates, and Thomas Andersson and Anna Maria Bengtsson from the Stockholm School of Economics provided valuable input in reviewing drafts of the original survey.

For their help in reviewing drafts and boosting our efforts, special thanks to our colleagues at Tufts including Sharon Green, Jean Williams, Patricia Dillon, Sarah Hammond Creighton, Gina Nortrom, James Noble, Jean Intoppa, Kate Atkinson, Anne Redfield, and Linfield Brown. Special thanks also to Lesley Byrne who worked on this project on a volunteer basis. The Center staff play an essential role in all our work. For all their work and constant reviving of our spirits, thanks to Pat Knibbs, Kris Kalil and Roni Dudley. Renée Micciche provided invaluable assistance, and Kate Philbin and Beth Smith contributed encouragement, expertise, and the most valuable of commodities, good judgement.

The preparation of this book has been a pleasure in the capable hands of Eric Valentine and Nita Romer of Greenwood Publishing Group.

Finally, our sincere thanks to a group of people who helped keep us on track and laughing. Rob Hollister, Director, Lincoln-Filene Center and Chair, Department of Urban and Environmental Policy, consistently helped us around the rough spots with his expressions of confidence. Tom Hellman, Vice President, Bristol-Myers Squibb, provided insight and encouragement, and acted as an interpreter until we became fluent in ''corporate.'' Members of our immediate families played key roles as well. Ed Flaherty did more than his share with grace, humor and patience while Margaret focused on this effort. Stower and Eliot Beals provided welcome diversions for Ann, and never tired of asking, ''Mom, do you have homework again tonight?''

Introduction

The global nature of business and intensified concern for the environmental impact of industry have focused attention on how companies manage their operations both inside and outside the United States. Using information gathered through extensive research, interviews, and tours of company facilities, as well as survey data, this book presents an in-depth look at how multinational companies manage environmental issues.

Although there are important differences in the practices of companies, the primary focus is not on good or bad practice as such. This book contains examples of each. Rather, the focus is on the range of management behaviors and the connection of these behaviors to outcomes. Often, what the managers themselves believe about the effectiveness of their programs is more important than any judgement made by others.

In a broader context, this book is about what companies can do to support sustainable development. The World Commission on Environment and Development defined sustainable development as a process of change in which the exploitation of resources, the direction of investments, the orientation of technological development, and the objectives of institutions are all in harmony and enhance both the current and future potential to meet human needs and aspirations.[1] Much of the discussion that follows attempts to define corporate environmental, health, and safety practice in a fashion that permits us to examine whether current practice will result in sustainable development. We conclude that there are cases in which companies are now doing more than they are required by regulation, but still they are not contributing to sustainable development.

We offer the following observations about corporate environment, health and safety programs:

- There are corporate practices and polices that we as environmental professionals feel are innovative and forward-thinking. There are situations, however, in which we feel practices and policies do not provide an adequate degree of protection. On the basis of our work, we believe that there is room for improvement in multinationals' environment, health, and safety practice and in the whole web of corporate decision making that ultimately affects practice in facilities.
- Many companies approach environmental issues from an explicitly home country, corporate, and short-term point of view. Practice will improve if boundaries are extended beyond the immediate region, organization, and time frame.
- Companies often fail to recognize and incorporate cultural aspects of environmental issues into their programs. Efficient and effective management of environment, health, and safety requires a systems approach that permits autonomy at the facility level, thus allowing for differences in local customs and circumstances to be accommodated.
- Educational institutions in general, but universities and graduate schools of business and engineering in particular, can play a key role in promoting corporate actions that are consistent with sustainable development.
- Environment, health, and safety (EHS) considerations should be integrated into the decision-making process at all levels. EHS should be a factor in determining not only how products are made but whether they are made.
- Champions for the environment are critical. A highly visible individual with resources to implement effective programs is an important factor in managing environmental issues on a global scale, and local champions are needed to translate broad directives into meaningful action.

This work, and indeed many of the efforts of the Center for Environmental Management, Tufts University (our sponsoring organization), are directed toward putting information into the hands of the decision makers with both public and private sector responsibilities. It is our hope that the findings discussed here will assist company environmental professionals with their increasingly challenging roles.

In addition, we believe that this information will be useful for the government decision maker and will facilitate the development of better, more effective, and competent environmental policies. Finally, the examples and discussion contained in this volume provide ideal teaching materials for the growing curricula in the arena of business and management as it relates to environmental implications.

Too often, important issues related to corporations and environment, health, and safety are debated in the news media in response to particular pollution incidents, such as the Exxon Valdez spill, or in anticipation of specific political decisions, such as the trade agreement between the United States and Mexico.[2] This public dialogue, while useful in raising consciousness about issues, generally fails to communicate the complexity of the issues and the tradeoffs that must be made by a vast array of decision makers. In the following chapters, we provide some of the history and context of corporate EHS issues, identify several of the approaches being used by companies to respond to increased demands for responsible actions, and explore the question of what drives good programs.

Chapter 1 lays the groundwork for many of the issues faced by multinationals in dealing with the environment. Chapter 2 discusses the approaches used to research this work as well as a brief profile of five case study companies. In conjunction with the case study research, a survey of 98 U.S.-based companies was conducted to gain additional information on EHS practices. An overview of the methodology and a profile of the respondents to the survey are discussed in Chapter 2, with findings presented throughout the book.

Chapter 3 covers information that is key to policy and organizational structure of EHS management. In Chapter 4, two case companies are described to highlight the issues raised in the previous chapter that are relevent to policies and structures. Chapter 5 presents working tools, or program components, that together constitute an EHS program. Chapter 6 examines program components through the perspective of three case companies: Oil and Gas, Household Products, and Pulp and Paper. International issues associated with EHS, specifically standardization of practice, are examined in Chapter 7. Following this discussion, Chapter 8 provides a snapshot of MNCs operating in Brazil and Mexico, with a specific discussion of two case company facilities in these countries. In Chapter 9 we analyze what drives good EHS programs; examples are drawn from the case companies. The conclusions are presented in Chapter 10, and an effort is made to operationalize sustainable development as it relates to corporations.

In recent years, we have increased our understanding of the effects of human activity on natural systems. We believe that understanding the institutions that convert natural resources to products and by-products is critical if the goals of sustainable development are to be met. We hope the material that follows helps to inform discussion, improve corporate practice, and provide directions for further inquiry on corporate environment, health, and safety management.

NOTES

1. World Commission on Environment and Development, *Our Common Future* (New York: Oxford University Press, 1987), pp. 8, 42.

2. See, for example, Jolie Solomon, "US Firms' Standards in Mexico Targeted," *Boston Globe*, February 13, 1991, p. 29.

Corporate
Responses
to
Environmental
Challenges

1

Multinational Corporations: Impacts and Challenges

INTRODUCTION

The decisions and conduct of international business simultaneously reflect and shape society's mores. This is increasingly apparent in the critical role that multinational corporations (MNCs) are assuming as major stakeholders in the environment. As agents for change, MNCs' decisions, along with those of government, will have a profound impact on the ability of society to make its development sustainable, "to ensure that it meets the needs of the present without compromising the ability of future generations to meet their own needs."[1]

A multinational corporation is "a network of enterprises that controls activities and assets in more than one state, and most often in three or more states."[2] Eugene Rostow and George Ball observe that the term *multinational corporation* is a misnomer; such companies have historically been incorporated in a single country and have had operations in several host countries.

"In legal personality, therefore, the multinational company is not multinational at all. Its authority derives from the law of a single state, coupled with rights granted to it by the host nations in which it is permitted to carry on business."[3] In strategy and philosophy, if not legal terms, many companies are beginning to undergo a transformation from multinational to global actors during the 1990s.[4]

This chapter provides a perspective on the resources commanded by MNCs and explores some of the forces external to the corporation that impact corporate decision making with respect to environment, health, and safety (EHS). The specific influences discussed in this chapter are as follows: factors that affect EHS management; regulatory requirements; host country development objectives and their possible conflicts with company objectives; and agendas of international organizations and nongovernment organizations. The chapter concludes by noting the formidable management challenges of implementing an EHS program in a multinational corporation.

A CHANGING STAGE

MNCs are undergoing considerable change as the globalization of the world's economy accelerates. Competitive pressures, mergers and acquisitions, pending economic unification of the European Community, rapid changes in Eastern Europe and the Pacific Rim, privatization in the emerging economies, acceleration of debt problems in Latin America and Africa, and the internationalization of finance markets are all factors affecting these changes. With few exceptions, companies are restructuring and consolidating positions in related industries rather than diversifying into others. Industries are becoming increasingly concentrated globally.

Since the recession in the early 1980s, many large MNCs have made significant staff reductions and, in an effort to modernize management, have pushed authority and responsibility further and further into division and facility structures.[5] Corporate EHS principles and practices are inevitably shaped by these larger forces and organizational trends.

Newsworthy environmental activities have prompted companies to evaluate their EHS strategies. The December 1984 release of methyl isocyanate from the Indian affiliate of the U.S.-based multinational Union Carbide in Bhopal, India, was more than a disastrous reminder that technological risk is present in our society. Early accounts of the release raised the question of whether the Union Carbide affiliate was operating with equivalent procedures, safeguards, and equipment to those at a comparable facility in the United States.[6] Because commensurate controls were not in place or operational, Bhopal has come to symbolize issues related to management responsibility and multinational operations in developing countries.

Along with the incident in Bhopal, corporations have cited increased public concern over other environmental matters, such as the Exxon Valdez oil spill in Alaska and the accumulation of scientific evidence regarding depletion of the stratospheric ozone layer, as important factors in motivating them to enhance their EHS efforts.

In addition, MNCs are reacting to external influences that have traditionally not weighed heavily in corporate decision making. Increased requirements for information disclosure, such as the Superfund Amendments and Reauthorization Act (SARA) Title III, Section 313,[7] in the United States, have encouraged a more informed and aggressive media, which simultaneously influences public opinion. Public concern for the environment is increasingly expressed through shareholder pressure and through consumer choice via the "green consumer" movement. In addition, nongovernment organizations are attempting to establish norms for corporate environmental actions. The impacts of these nontraditional sources of persuasion on environmental decision making may become significant in the long run.

Corporations are beginning to take actions in three areas: acquiring more information about their operations worldwide, particularly their joint ventures and subsidiaries; developing systems to minimize the chances of low probability/high consequence events occurring, and to minimize the damage to people and property if such events do occur; and establishing comparable EHS activities in the many countries in which they operate.

MAGNITUDE AND SECTORS OF OPERATIONS

According to the United Nations Centre on Transnational Corporations (UNCTC), activities of multinational companies "affect at least one quarter of the world's productive assets; 70 percent of the products in international trade; 80 percent of the world's land cultivated for export crops, and the major share of the world's technological innovations."[8]

In addition to the sheer size of their operations, MNCs tend to dominate environmentally sensitive sectors including minerals, oil and gas development, agribusiness, and chemicals. It is the concentration in vulnerable industrial sectors that has drawn worldwide attention to MNC operations.[9]

Reports on the trade in environmentally hazardous or unsafe products abound, with DDT and asbestos as well-known examples. Less publicized but equally distressing is the sale by Velsicol Chemical Company of an unapproved pesticide to Egypt, Indonesia, and South Vietnam that resulted in the death of one person, the poisoning of 65 persons who were field workers, and the deaths of several hundred water buffalo in Egypt alone.[10] These reports continue to cause apprehension in both developed and developing countries.

There is much controversy regarding MNCs adhering to different standards of environmental performance depending on where in the world they operate a facility. Do companies from industrialized and regulated countries take advantage of different and presumably lower standards in developing regions? Are companies locating dangerous or polluting activities in areas where EHS regulations are weak? Jeffrey Leonard found that some industries, including primary metals processing, asbestos, and some chemical manufacturing industries, such as those manufacturing benzidine-based dyes and trichlorophenol, were indeed leaving the United States in response to stricter environmental regulations.[11] The Border Industrialization Program, originally created in 1965 to entice foreign investment, has resulted in 1,800 factories operating in the northern border towns of Mexico. These "maquiladoras" are an example of a developing region, desperate for jobs, confronted with the conflicts and contradictions of developing in a sustainable manner.[12]

The other side of the coin, equally well documented, is that in general MNCs have a better record of performance in developing countries than do local enterprises.[13] The reasons advanced for this contrast are as follows:

1. The MNCs are more visible and thus subject to intense scrutiny by host governments. It may be more politically acceptable and less risky for host governments to enforce environmental standards against MNCs as a sign to the public that the government is serious about environmental concerns.[14]
2. Home country demands are more stringent—from government, stockholders, press, and environmental nongovernment organizations.
3. Because of their superior assets, MNCs are better able to incorporate improved technology and training.
4. Because these changes have already been incorporated in home country operations, it is cost-effective to replicate them in host country facilities.[15]

RESOURCES AND STRUCTURE

MNCs are by no means monolithic. They very often consist of a number of separate companies in vastly different businesses with substantial internal differences in EHS problems, management styles, and degrees of autonomy. Personnel, technological innovation, and financial investments, can, of course, make a large difference not only to worldwide environment, health, and safety and to the reputations of firms, but also to the realistic chances for sustainable economic development.

The pattern of mergers, acquisitions, and divestitures that characterizes U.S.-based MNCs also means that comparable manufacturing operations, such as electroplating, could be conducted by different units or divisions within the same corporation, with different management approaches and different outcomes for the environment. This variation in approach and outcome raises concerns associated with internal corporate standards as well as the development, implementation, and enforcement of global standards.

Other complicating structural conditions include joint ventures, with both government or private partners; mixed ownership-management situations, in which companies manage government-owned facilities under contract; and absentee ownership arrangements. Determining whose EHS rules apply in these circumstances is critical. In many companies this is increasingly becoming a matter that is negotiated at the time the relationships are formed, although substantial ambiguity may continue in practice.

These complexities present both legal and management challenges. The relatively good reputations enjoyed by the MNCs under these circumstances may be a tribute to their management acumen, especially with respect to health and safety matters. Corporate EHS staff are not complacent; one commented, "Sometimes you can get by on the quantitative side in the environment area because of luck." Another corporate environmental manager observed, "We know we're out of compliance somewhere in the world all of the time."

MOTIVATING FACTORS

Despite the wide differences in products, technology, and business strategy, corporations have many common practices. The fact that environment, health, and safety are regulated extensively in the United States and to varying degrees in most countries undoubtedly accounts for some of the commonality in management approach. Generally, corporations rely on a relatively small number of outside sources for EHS management information, including consultants and industry associations, which may also account for the degree of commonality in practice.

The terminology for the content of corporate programs varies from one company to the next, and some differences in word choice are conscious. The generic term of *environmental audit* is used by many companies, but one company, for example, expresses a strong preference for the term *assessment*, explaining that it creates a less judgmental interaction between corporate headquarters and facilities.

While there are many common program components, there are also important differences. These distinctions are apparent in terms of the way in which companies view themselves and their relationship to the countries in which they operate. An executive at General Motors in Brazil characterizes the firm as "a totally American company, no matter where in the world we operate." In contrast, other companies strive for a more global identity. Some companies have addressed the issue of global identity by denationalizing their names; for example, British Petroleum is now BP, and Badische-Analin Soda-Fabrik, a German firm, was renamed BASF.[16] There are also industries with strong national identities that have recently become active in overseas investments. PetroBras and Pemex, the state-owned Brazilian and Mexican petroleum industries, are examples.

Differences in the country context can also have an impact on EHS priorities. The following factors tend to influence the internal decision making of companies with respect to environment, health and safety.

1. *Degree of development*. MNCs face different challenges depending on the degree of development and industrialization of the country or region in which they are operating. For example, if a country is facing fundamental infrastructure development and is struggling with sewage disposal, delivering potable water supplies, and eradicating infectious disease, this will influence the manner in which a corporation deals with, for example, health issues among the work force. Similarly, if there are frequent interruptions in the supply of electricity, it will have an impact on technology choices for production and environmental protection.

2. *Geographic location*. Different priorities within EHS programs may be attributable to facility location. For example, the topography of Mexico City, like that of Los Angeles, has a tremendous impact on the air quality in the metropolitan region. As a consequence, MNC environmental personnel in Mexico City are inundated with air monitoring concerns and spend a proportionately greater amount of their time on air pollution issues than do their counterparts in other locations.

3. *Public concern*. Community action can catalyze industry efforts to make improvements that are related to environment, health, and safety. Noise and odors emanating from plants are common environmental issues addressed in response to public pressure by facilities in Brazil and Mexico. In contrast, U.S. headquarters and facilities spent considerable effort preparing information for and anticipating

public reaction to the disclosure of information on selected toxic releases, as required by U.S. law (SARA Title III, Section 313). Public concern in France and Great Britain can be attributed to Green party politics.

4. *Security issues*. Security in developing countries can be important, especially because manufacturing or assembly facilities can be located in heavily urbanized or urbanizing areas and surrounded by large numbers of newly settled squatters. Threats of sabotage from a discontented work force, from the neighboring community, or from general political instability can focus attention on lockout/tagout procedures, grounds security, and accelerating capital expenditures to computerize and mechanize vulnerable industrial processes. Some companies report situations in which travel by U.S. executives to foreign operations has been either curtailed or eliminated due to actual or perceived political turbulence in particular host countries. Other companies use private agencies to prepare security profiles of specific locations and potential threats to MNC personnel traveling to these sites.

5. *Regulatory context*. The EHS regulatory programs in place in the home country and the host country can have an impact on a company's performance. Critical issues involve not only what regulations are in place but also the nature of the relationship between the government and the regulated community[17] and the degree to which laws and regulations are enforced.

GOVERNMENTS AND NONGOVERNMENT ORGANIZATIONS

Multinational companies do not act in a vacuum. In addition to stockholders and stakeholders (employees and customers), governments at all levels, both home and host, are critically important shapers of MNC behavior. Beyond establishing and enforcing regulations on environment, health, and safety, depending on the country, government policies and actions can and do affect and control pricing, interest rates, subsidies, hiring practices, wages, tariff and trade practices, technological innovation, education and training of the work force, and the full range of matters vital to MNC operations.

In developing countries, loan conditions imposed by the World Bank and the International Monetary Fund have profound impacts on governments and therefore on MNC behavior.[18] Governments are also influenced by a wide range of other institutions, whose effects are indirectly felt by MNCs. In developing countries, these include the various specialized bodies of the United Nations system, whose economic and social models, training perspectives, and guidelines all influence the thinking of the various government ministries that interact with MNCs.

In addition to the United Nations–sponsored activities, there is usually a substantial number of bilateral development assistance bodies operating. These institutions, such as the U.S. Agency for International Development, often respond to the larger global political interests of their home country, but nevertheless they are a factor in the host country government operations. As providers of money and technical assistance, they often determine the infrastructure in which MNCs must operate. These assistance agencies are pressed by advocates at home to condition their ventures on environmental protection and to negotiate with the host government over these issues. Host government coordination of the activity is usually quite modest in scope, and cooperation by one or more outside bodies is usually actively discouraged by the government to avoid being outnumbered in negotiations. Recipient governments also actively resist "strings" on foreign aid, such as environmental conditionality, in order to preserve options.

The establishment of the UNCTC reflected the differences in priorities between some multinationals and developing countries. In 1977, a working group began drafting a Code of Conduct for Transnational Corporations[19] to address these issues; however, differences in approach and interests between developed and developing countries have prevented the group from forging consensus. The Code of Conduct remains in draft form pending resolution of several issues, particularly matters related to proprietary technology.

Also operating in the developing country natural resource area are private voluntary agencies including church groups; various private food and development aid groups such as Oxfam; and both developed and developing country nongovernment organizations with environment missions, such as Greenpeace, the World Wildlife Fund, and the Nature Conservancy. The Environment Liaison Centre with headquarters in Nairobi, for example, lists more than 4,000 active member organizations that are predominately from less developed countries. All these groups have both technical and political impacts on host governments, some of which are felt in government dealings with MNCs.

REGULATORY RELATIONSHIPS

In the early 1970s and accelerating into the 1980s, most developing country host governments enacted EHS legislation. However, even in the early 1980s a spokesman from a large MNC operating in several developing countries said in public "that his company, in the majority of cases, 'brought the governments along' toward accepting higher environment standards, often in the face of their short-term economic arguments."[20]

In Brazil and Mexico, two of the five[21] countries visited as part of the research for this work, environmental laws were largely based on similar legislation in the industrialized world, specifically the United States.[22] Individuals in Mexico, whether they were from industry, government, nongovernment organizations, or academia, were in agreement that the enforcement component that makes the U.S.-style regulatory system function is not in place. A government official observed that "MNCs are definitely not meeting the standards" and later noted that "if we try to force compliance, a good Mexican lawyer can delay things for two or three years."

Employees from the Mexican environmental regulatory agency lamented the fact that so little equipment is available to conduct monitoring functions, essentially foreclosing opportunities to collect and maintain a verifiable data base on polluters. Industry spokespersons agreed with this assertion; however, they noted that the situation in Mexico is further exacerbated by the fact that regulatory agency personnel are not adequately trained to analyze data even if equipment were available for broader use.

More recently, Mexican government officials have begun to question whether industrialized country perspectives on environmental control are relevant to their situation. As a result, techniques for analysis of risks from chemicals developed in countries belonging to the Organisation for Economic Cooperation and Development (OECD) have not been adopted quickly. Developing countries often find themselves with more pressing problems than cancer risk; problems such as delivering a potable water supply affect populations at younger ages.

A Mexican government official noted the millions of dollars spent by the United States for environmental protection and the results reported in the international media and is searching for a better way to approach the problems of environment and development in his country. Instead of simply estimating and trying to mitigate environmental impacts, for instance, as is done elsewhere, the Mexican government now wants to be apprised of potential implications: The official asked, "beyond jobs, for a given cost of environmental impact, what will be the benefit to the Mexican people?"

Frequently, in less developed countries, as support for various environmental goals gains ground both among their own middle class and in the international press, regulations are added to the books but resources are not available to enforce them. MNCs are thus often in a position to operate flexibly just at compliance levels or some degree above or below them, according to company policy or plant manager outlook.

The issue of inadequate enforcement resources is not unique to developing countries; it has been a recurring concern in the United States and Europe among environmental groups. Frequently, company officials conveyed a different attitude toward compliance with regulations in the United States than they did in Brazil and Mexico. The specific terminology of goals for the U.S. operations varied, but the general target was "zero violations." In contrast, we asked the EHS manager of a large U.S.-based corporation about the firm's environmental posture in Brazil, where parts of the U.S. regulations have been adopted. The response was:

> But in Brazil there is no enforcement. Here is the situation in Brazil: By regulation the company is required to install secondary treatment for its sanitary discharge, but right next to the plant is the out-fall from a 2,000-person apartment building where the stuff comes out untreated. What does the company do? We eventually put in the secondary treatment, but at a much slower schedule than we would if there were enforcement in the country.

The issue of water treatment was raised by several companies having operations in Brazil and Mexico. Companies noted that in addition to treating effluent, they frequently have to treat incoming water arriving at the facility from upstream that is too contaminated for process use.

ENVIRONMENTAL TRENDS

MNCs and their trade groups have been able over the years to monitor and, to some extent, influence the development of national and local laws in the countries in which they operate. At the same time, an increasingly large body of regional and international environmental law has been developing, some of which is vital to MNCs. MNCs in the mining and oil sectors, for example, were participants in the 1970s in international discussions surrounding mineral exploitation aspects of both the Law of the Sea Treaty and a convention to regulate, explore, and exploit minerals in the Antarctic. More recently, the chemical MNCs were key participants in the discussions surrounding the control of chlorofluorocarbon (CFC) emissions in the development of the Montreal Protocol[23] and have actively influenced the outcome of government and international agency discussions on the transboundary movement of pesticides.

These developments in international environmental law, which will probably include some formal efforts to limit industrial contributions to the buildup of greenhouse gases in the atmosphere, have forced some MNCs to more seriously monitor and otherwise participate in deliberations of various international organizations, including the United Nations Environment Programme (UNEP) and the annual meeting of the United Nations General Assembly. In addition, the OECD and the European Economic Community have promulgated guidelines that, because of the prestige of these bodies and the careful way in which their positions are developed, are taken seriously by the MNCs.[24]

At the same time, industry itself has been developing voluntary guidelines that apply to environmental issues. Voluntary codes of conduct are an important aspect of MNC operations because they avoid the sovereignty issues that are associated with regulating MNC actions. Such guidelines are discussed throughout this book, and a brief context is included here.

In 1984, the International Chamber of Commerce (ICC), in conjunction with UNEP, held a conference called "The World Industry Conference on Environmental Management." Among the major recommendations of this conference was the following:

> To strengthen the anticipatory and preventative approach to environmental management within industry, each line manager from the chief executive down should also think of him or herself as an environmental manager. Clear accountability for environmental performance should accompany managerial responsibility in each case.[25]

In the same year, the World Resources Institute convened a panel of business leaders and other experts to discuss the roles of MNCs in improving environmental cooperation with developing countries. Among the conclusions of this discussion were the following recommendations: (1) Most important, it should be explicit corporate policy to comply fully with the host country's laws and regulations; and (2) Policy must be set at the highest level of management and demonstrate top management's real and continued commitment.[26]

Examples of industry guidelines include those adopted by the ICC[27] and also by a number of sectoral trade groups, such as the International Petroleum Industry Environment Conservation Association (IPIECA). The following is a sample of the issues addressed in the ICC's guidelines:

> Industry has particular environmental responsibilities in terms of such factors as plant location and design, process selection and product design, environmental pollution, harmful radiation, vibration and noise controls, waste disposal, occupational health and safety aspects and long-range planning.

> Each company's management should promote among its employees at all levels an individual sense of environmental responsibility and should educate and encourage them to be alert to potential sources of pollution and to sound resource conservation measures within their operations.[28]

A more recent MNC response to environmental concerns is the formation by a group of U.S. multinationals of a Global Environmental Management Initiative (GEMI). The goals of the GEMI are as follows: (1) to stimulate, assemble, and promote worldwide critical thinking on environmental management; (2) to improve the environmental performance of businesses worldwide through example and leadership; (3) to promote a worldwide business ethic for environmental management and sustainable development; (4) to enhance the dialogue between business and its interested publics such as nongovernmental organizations, governments, and academia; and (5) to forge partnerships around the world to encourage similar efforts in other countries.[29]

The Chemical Manufacturers Association (CMA), a U.S. organization, has developed a set of what it calls "Guiding Principles for Responsible Care of Chemicals." Companies must pledge to operate according to the principles as a condition of membership in the CMA. Four of the ten principles are presented as follows: (1) to make health, safety, and environmental considerations a priority in our planning for all existing and new products and processes; (2) to counsel customers on the safe use, transportation, and disposal of chemical products; (3) to operate our plants and facilities in a manner that protects the environment and the health and safety of our employees and the public; and (4) to extend knowledge by conducting or supporting research on the health, safety, and environmental effects of our products, processes, and waste materials.[30]

A second World Industry Conference on Environmental Management (WICEM II) was held in April 1991 and was organized by the ICC. An effort was made to assess progress since the first WICEM[31] and to create momentum for elevating consideration of environmental issues in all corporate decisions. Industry participants were encouraged to sign the 16-point Business Charter for Sustainable Development, which was prepared by the ICC. The first principle, on corporate priority, commits companies "To recognize environmental management as among the highest corporate priorities and as a key determinant to sustainable development, to establish policies, programmes and practices for conducting operations in an environmentally sound manner."[32]

In the industrialized nations, there is a growing claim by MNCs and an increasing government conviction that industry's approach of developing and applying "end-of-pipe solutions" to environmental problems (since the early 1970s) has put companies on the defensive and forced them into a "compliance trap," away from more cost-effective pollution prevention and waste minimization approaches. Corporations such as Monsanto and Du Pont have become ardent advocates of pollution prevention and have characterized this approach as critical to a new environmental stewardship that responsible companies must practice.[33,34] Unilateral actions are being taken by some MNCs to reduce wastes and emissions well below levels required by government. For example, 3M hopes to reduce emissions of hazardous materials by 90 percent by the year 2000.[35] This pattern, if followed more generally, could represent an important advance toward sustainable development.

TRAGEDY AS A CATALYST

A benchmark for sharpening both public and private attention to corporate EHS responsibility was the 1984 tragedy in Bhopal, India. Union Carbide, the U.S.-based multinational, had 50.9 percent ownership in a pesticide plant that was operated by the subsidiary, Union Carbide India, Ltd. Although several investigations of the incident have been made, the issues raised there were manifold and controversy still surrounds the circumstances.

It is clear, however, that substantial liability, even under a less than 100 percent ownership situation, can accrue to the home country firm. Corporations were quick to see the implications, and as one corporate staff member said, "The day after Bhopal, chief executive officers all over the United States were calling in their lawyers and environment managers and asking, 'Could that happen to us?' "

A critical component of the aftermath of Bhopal was recognition of the potential risk to the surrounding community from industrial activities and, more important, the need for the surrounding community to know the magnitude of the risk and, in cooperation with the firm and local authorities, to have a coherent emergency plan. In the United States, this resulted in federal legislation that affects inhabitants of communities surrounding plants, establishing formal "right-to-know" procedures regarding hazards associated with the compounds in use, requiring publication of inventories of toxic compounds, and mandating emergency planning.[36] Many states and localities have also passed complementary legislation. Several company people who were interviewed expressed the view that it is just a matter of time before some variation of the "community right-to-know" policy will become law in most countries where they have operations.

It is worth noting that although the Bhopal incident was an epic tragedy the reverberations of which were felt worldwide, between 1921 and 1984, 25 major industrial accidents resulted in a total of 8,432 fatalities, not counting the incident in Bhopal.[37] Fatalities from selected incidents are detailed in Table 1.1.

Table 1.1
Selected Industrial Incidents Causing Fatalities

YEAR	ACCIDENT	SITE	FATALITIES
1921	Explosion in chemical plant	Oppau, Germany	561
1942	Coal dust explosion	Honkeiko Colliery, China	1572
1947	Fertilizer ship explosion	Texas City, USA	562
1950s	Seafood contamination	Minamata Bay, Japan	439
1956	Dynamite truck explosion	Cali, Colombia	1100
1950s	Foodstuff contamination	Turkey	400
1970	Natural gas explosion	Osaka, Japan	92
1972	Foodstuff contamination	Iraq	459
1974	Chemical plant explosion	Flixborough, UK	28
1975	Mine explosion	Chasnala, India	431
1976	Dioxin leak	Seveso, Italy	0
1978	Road accident/gas explosion	Xilatopec, Mexico	100
1978	Natural gas explosion	Hulmanguille, Mexico	58
1978	Transport accident/propylene	Los Alfaques, Spain	216
1979	Bio/chem warfare plant accident	Novosibirsk, USSR	300
1981	Rail accident/chlorine leak	Montana, Mexico	29
1982	Tank explosion	Caracas, Venezuela	101
1983	Food contamination	Madrid, Spain	340
1984	Natural gas explosion	Mexico City, Mexico	452
1984	Pipeline explosion	São Paulo, Brazil	508
1984	MIC gas leak	Bhopal, India	2500
1984	Natural gas explosion	Garhi Dhoda, Pakistan	60
1986	Nuclear plant explosion	Chernobyl, USSR	*250
1987	Methyl alcohol related	Guangxi, China	55

Sources: B. Bowonder, Jeanne X. Kasperson, and Roger E. Kasperson, "Avoiding Future Bhopals," *Environment* (27) 7 (1985): 8; United Nations Environment Programme, *Environmental Data Report* (Boston: Basil Blackwell, 1989), p. 403–404.
*Updated figure quoted from Moscow News in David Remick, "Soviets Report 250 Deaths Occurred at Chernobyl; Official Toll Following 1986 Nuclear Accident in the Ukraine Had Been Only 31," *Washington Post*, November 9, 1989, p. a70.

Incidents at Seveso, Chernobyl, and the Sandoz Rhine plant,[38] as well as other less dramatic episodes, have brought about an internationally increased sensitivity to problems of technological risk. From the point of view of MNCs, this sensitivity can often translate to issues of facility siting, increased worker concern about chemical safety, and a host of other issues, including greater government oversight. For example, the 1976 release of dioxin at a Hoffmann La-Roche plant in Seveso, Italy, caused no fatalities,[39] but it was the catalyst for the European Community's development of measures on risk evaluation, communication, and emergency planning, commonly called the Seveso Directive.[40] In addition, some developing countries are beginning to use risk zoning in city planning in an effort to control the location of various industrial facilities.[41]

MANAGEMENT IMPLICATIONS

MNCs cannot achieve, or go beyond, compliance without a concerted management effort. In finance, manufacturing, distribution, and marketing, the ability to provide incentives, hire, train, promote, and the like has been a hallmark of successful business. Recent modifications in organizational structure, however, are placing considerable strain on management skills in the EHS area. EHS management is an evolving field, requiring special technical skills together with superior selling and communications attributes. As managements decentralize and develop approaches for giving greater autonomy to local managers, methods need to be developed to assure headquarters of cost-effective information flow and to monitor compliance. At the same time, EHS managers at headquarters need to justify the expense of their programs in the absence of more concrete cost-benefit measures. As an EHS plant manager in Mexico said: "It's difficult to prove the value of a job well done in the EHS area because if you're doing a good job, nobody hears anything about your activities. It's only when something goes wrong that all the noise starts."

NOTES

1. World Commission on Environment and Development, *Our Common Future* (New York: Oxford University Press, 1987), p. 8.

2. George Modelski, ed., *Transnational Corporations and World Order* (San Francisco: W. H. Freeman and Company, 1979) p. 1.

3. Eugene V. Rostow and George W. Ball, "The Genesis of the Multinational Corporation," in *Global Companies: The Political Economy of World Business*. George W. Ball, ed. (Englewood Cliffs, N.J.: Prentice Hall, 1975), p. 4.

4. Kenichi Ohmae, "Planting for a Global Harvest," *Harvard Business Review* (67) 4 (1989): 136–145.

5. Philip R. Harris, *Management in Transition: Transforming Managerial Practices and Organizational Strategies for a New Work Culture* (San Francisco: Jossey-Bass, 1985) pp. 1–29.

6. "Union Carbide Fights for Its Life," *Business Week*, December 24, 1984, pp. 53–56.

7. The Emergency Planning and Community Right-to-Know Act of 1986, P.L. 99–499, was enacted as Title III of the Superfund Amendments and Reauthorization Act. Section 313 requires certain facilities to report releases of toxic chemicals to all media (air, water, land) and further mandates that the Environmental Protection Agency must create a national inventory based on the data and make it available to the public. See U.S. Environmental Protection Agency, "Title III Fact Sheet, Emergency Planning and Community Right-to-Know," August 1988.

8. United Nations Centre on Transnational Corporations, "Criteria for Sustainable Development Management of Transnational Corporations," Background paper No. 1, Expert Group Meeting, December 18–20, 1989, New York, p. i.

9. Economic and Social Commission for Asia and the Pacific, *Transnational Corporations and Environmental Management in Selected Asian and Pacific Developing Countries* ESCAP/UNCTC Publication Series B, No. 13 (Bangkok: ESCAP, 1988), pp. 3–4.

10. Rashid A. Shaikh, "The Dilemmas of Advanced Technology for the Third World," *Technology Review* 89 (1986): 57.

11. H. Jeffrey Leonard, "Confronting Industrial Pollution in Rapidly Industrializing Countries: Myths, Pitfalls, and Opportunities," *Ecology Law Quarterly* 12 (1985): 779–783.

12. Joseph LaDou, "Deadly Migration," *Technology Review*, July 1991, p. 49.

13. ESCAP, *Transnational Corporation and Environmental Management*, p. 7.

14. Comment made by a corporate executive during an interview with a case company.

15. For detailed discussion of these issues, see generally UNCTC, United Nations Centre on Transnational Corporations, *Environmental Aspects of the Activities of Transnational Corporations: A Survey*, ST/CTC/SS, UN Pub. No. E85.11.A.11, 1985, N.Y., N.Y.

16. Raymond Vernon, *The Economic and Political Consequences of Multinational Enterprise: An Anthology* (Boston: Division of Research, Harvard Business School, 1972), p. 121.

17. See Norman Beecher and Ann Rappaport, "Hazardous Waste Management Policies Overseas," *Chemical Engineering Progress*, May 1990, pp. 30–39.

18. See, for example, "Environmental Requirements of the World Bank," *The Environmental Professional* 7 (1985): 205–212.

19. Draft United Nations Code of Conduct on Transnational Corporations, reproduced in *International Legal Materials* 23 (1984): 626–640.

20. J. Gary Taylor, "Environmental Planning in the Context of Development Investment," Report of the First Talloires Seminar on International Environmental Issues (Medford, Mass.: Tufts University Department of Urban and Environmental Policy, 1983), p. 13.

21. The countries hosting the foreign operations for this research include Brazil, Mexico, Great Britain, France, and Canada.

22. Additional information on EHS regulations in Brazil and Mexico is included in Chapter 8, which provides an in-depth view of selected aspects of study companies.

23. United Nations, Protocol on Substances That Deplete the Ozone Layer, Montreal, September 16, 1987, reproduced in *International Legal Materials* 26 (1987): 1541–1561.

24. See Organization for Economic Cooperation and Development, "Guidelines for Multinational Enterprises," reproduced in *International Legal Materials* 15 (1976): 969–976.

25. Alex Trisoglio and Kerry ten Kate, "From WICEM to WICEM II: A Report to Assess Progress in the Implementation of the WICEM Recommendations," United Nations Environment Programme Industry and Environment Office, Paris, France, March 1991, p. 3.

26. World Resources Institute, "Improving Environmental Cooperation: The Roles of Multinational Corporations and Developing Countries," Report of a panel of business leaders and other experts convened by the World Resources Institute (Washington, D.C.: World Resources Institute, 1984), p. 36.

27. International Chamber of Commerce, *Environmental Guidelines for World Industry*, Pub. No. 435 (Paris: ICC, 1990).

28. ICC, *Environmental Guidelines*, pp. 8–9.

29. George Carpenter, "First Volume," *GEMI-News*, September 1990, p. 1.

30. Chemical Manufacturers Association, "Responsible Care, A Public Commitment, Questions and Answers about Responsible Care," April 1991.

31. See Trisoglio and ten Kate, "From WICEM to WICEM II."

32. ICC, "The Business Charter for Sustainable Development, Principles for Environmental Management," Publication 210/356 A (Paris: ICC, April 1991).

33. T. H. Lafferre, Speech to American Institute of Chemical Engineers, Conference on Waste Minimization, Washington, D.C., December 4, 1989.

34. See also E. S. Woolard, "Corporate Environmentalism," Remarks by the chairman, Du Pont, before the American Chamber of Commerce, London, May 4, 1989.

35. "3M Announces Plan to Cut Hazardous Releases by 90 Percent, Emphasize Pollution Prevention," *Environment Reporter*, June 16, 1989, p. 441.

36. Emergency Planning and Community Right-to-Know Act of 1986, P.L. 99–499, Title III.

37. B. Bowonder, Jeanne X. Kasperson, and Roger E. Kasperson, "Avoiding Future Bhopals," *Environment* (27) 7 (1985): 8.

38. See Linda C. Durkee, "Risk Communication and the Rhine River," *International Environment Reporter* 12 (October 11, 1989) Part II.

39. "Union Carbide Fights for Its Life," *Business Week*, p. 54.

40. See Michael S. Baram, "Risk Communication Law and Implementation Issues in the United States and the European Community," in *Corporate Disclosure of Environmental Risks: U.S. and European Law*, eds. Michael S. Baram and Daniel G. Partan (Salem, N.H.: Butterworth Legal Publishers, 1990), pp. 65–88.

41. J. Gary Taylor, "Managing Environmental Risk in Newly Industrializing Countries," Report of the Fourth Talloires Seminar on International Environmental Issues (Medford, Mass.: Tufts University, 1987), p. 11.

2

Case Study Protocol
and Survey Method

STUDY FOCUS

The material for this book is based on two sets of research activities. Initially, five in-depth case studies of U.S.-based multinational corporations were conducted to understand the complex forces both inside and outside the corporation that help shape the EHS programs in different locations around the world. Second, a survey was developed to test whether case study findings on corporate practices applied to a larger population.

CASES

The five MNCs selected for detailed analysis are identified in this book by their major businesses described as follows:

- *Oil and Gas* is a diversified company whose businesses include oil and gas production and chemical manufacture.
- *Household Products* has businesses that include manufacture of personal care items and laundry and cleaning products.
- *Chemicals* manufactures agricultural and specialty chemicals.
- *Pulp and Paper* grows and harvests timber and manufactures, distributes, and sells a wide variety of forest and paper products.
- *Instruments* manufactures process control instrumentation for a variety of industries, including chemicals, paper, and food.

COMPANY SPECIFICS

In 1989, these five companies had, in the aggregate, $48.3 billion in sales; 211,100 employees; and 225 non-U.S. subsidiaries.[1] Table 2.1 presents 1988

Table 2.1
Case Company Financial Indicators

	1988 Total Assets[a]	1988 Net Income[b]	# of Foreign Subsidiaries
Chemicals	5,310	233	128
Household Products	14,820	1,020	11
Instruments	472	9.8	14
Oil and Gas	20,747	303	17
Pulp and Paper	7,851	515	25

[a]Reported in millions of 1988 dollars
[b]Data obtained from corporations' 10-K filings

assets, income, and number of foreign facilities registered with the Securities and Exchange Commission for each of the five case companies.[2]

Cases were designed to examine three general types of issues. First, how does the corporation address EHS issues? In general, are there differences between programs and practices in the home country and in non-U.S. facilities? If there are differences, how are decisions made as to how much EHS protection is delivered in a particular location?

Second, does information on environment, health, and safety get "lost in the translation" from headquarters to facilities and vice versa? If so, how does the loss occur, and is it greater from headquarters to a home country facility than from headquarters to a foreign facility?

Third, how is the corporation organized and staffed to address EHS challenges? Why is it organized this way? How do organizational structure and formal and informal relationships within the corporation affect EHS programs?

PROTOCOL

Interviews were conducted at each case company's corporate headquarters, a domestic manufacturing facility, and at one or more of the company's overseas manufacturing facilities. The overseas locations included Canada (Pulp and Paper); Brazil (Oil and Gas); Mexico (Consumer Products and Oil and Gas); Great Britain (Instruments); and France (Chemicals).

Since it was believed that regulatory, economic, and social conditions in the host country were likely to have an effect on a company's EHS programs, the host countries were held constant in two cases in order to strengthen the observed outcomes. As part of the research, we met with government officials of host countries, U.S. consulate officials, representatives of nongovernment organizations, trade associations, media spokespersons, and academics.

There were also opportunities to tour facilities on companies other than the case study companies. This provided an additional perspective on how other operations are run in a similar setting. Therefore, it is possible that when a reference is made (for example) to a plant in Mexico, the information is from a company other than one of the five case study companies.

The original objective was to match as closely as possible, in terms of age and product, the facilities in and outside the United States for each company. It was anticipated that comparability would be increased if pairs of facilities of similar age producing similar products were identified. We were successful in identifying matches and conducting interviews at facilities in and outside the United States that produced similar products; however, age matches were generally not available within the case companies.

The case study approach was used because a rich and detailed discussion of a small set of examples can constitute a valuable contribution to understanding the complex decision making that lies behind corporate action in the EHS area. We offered each prospective study company the assurance that we would present its case without reference to the company name and that we would work with the company to ensure that other information we might include would not be an obvious clue to the corporation's identity.

In selecting cases, we did not seek companies whose practice we thought would be "representative" in the statistical sense. Statistical significance is not a possibility with such a small sample; instead, our cases are used for an analytical process called explanation building.[3]

STUDY PROPOSITIONS

In examining the case companies, we tried to unravel why some companies take a particular set of approaches to environment, health, and safety, and why others select very different approaches that may have different implications for environmental protection and human health.

An early step in the research was identifying and articulating a set of propositions, or causal statements. Case companies were selected to permit the researchers to conclude, once the data was collected and analyzed, that propositions are supported or not supported by the evidence, or that propositions should be modified.

We formulated the initial propositions to examine selected areas of interest as reported in the literature. Several prior research efforts have addressed issues related to corporations and the environment, and they served as building blocks for our effort.

Roger Kasperson et al. used case studies to examine the closely related field of hazard management and identified a group of exogenous variables—including external regulation, liability and insurance costs, and public scrutiny—all of which were seen as important in "forcing reluctant corporations to upgrade their hazard management programs." Endogenous variables identified by Kasperson et al.

include profitability, commitment of high-level management, age of facility, and degree of hazard associated with the product. In addition, Kasperson et al. noted that it is commonly asserted that much of the improper activity occurs within small firms; however, no small firms participated in their study.[4]

In the mid-1970s, Thomas Gladwin performed a comprehensive analysis of corporate consideration of environmental issues at the project level. Using corporate interviews as a primary source, he found that the extent to which environmental issues are addressed (extent of "ecological incorporation") in MNC planning is a function of (1) the nature of the project, (2) the organizational context of the planning process, and (3) the public policy context of the firm and the project being planned. Among his findings Gladwin determined that top management support and extent of external scanning were independently but moderately positively correlated with extent of ecological incorporation, and that the relationship between profitability and incorporation was genuine but weak.[5]

Jeffrey Leonard examined the hypothesis that U.S. firms tend to locate overseas in areas where EHS regulatory systems are weak, and he found the hypothesis to be supported in only a small number of industries.[6] Leonard's findings are supported by the UNCTC[7] and by Allen White.[8]

In 1984 the International Labour Organisation published a study of eight MNCs in which the issue of home country and host country health and safety standards was explored. Interviews were conducted at corporate headquarters, in another industrialized country, and, when possible, in a developing country. The study concluded that home country regulatory standards form the basis for health and safety programs in host countries; however, adaptation occurred to reflect variations in host country requirements. In some cases, corporation-wide standards were developed and implemented. The study also concluded that "it is still true that safety and health standards in the Third World countries are less developed than in the industrialized countries."[9]

Patricia Dillon and Kurt Fischer studied 15 companies with varying environmental records and reputations in an effort to determine what internal strategies and programs the companies use, why they take these approaches, and how they achieve environmental goals. Features of successful management programs were identified to include various aspects of organizational structure, environmental planning, and management controls. Motivators for strong programs included costs and government requirements; company culture, tradition, responsibility, and values, particularly a concern for protecting the "good name" of the company; and strong leadership.[10]

In a 1989 study, Mitchell Koza et al. conducted interviews and examined nine European companies' policies for environmental protection. Four issues emerged that Koza et al. believe to be important: (1) The successful development and implementation of environmental practices often require a company "champion," a manager to spearhead environmental initiative; (2) Companies organized on a product line basis may experience difficulties implementing country- or region-wide environmental policy; (3) Environmental partnerships among businesses, advocacy groups, and public officials appear to be a viable response to environmental issues; and (4) Aggressive response to environmental issues may lead companies to become relatively visible in the press and to the advocacy community.[11]

After considering the studies just mentioned, we formulated the following set of propositions to further test conclusions drawn by others and to examine challenges associated with global operations.

1. *There will be differences in the effectiveness of the companies' EHS practices that can be explained by the type of business in which each company is engaged.* For example, we expected to see similar practices among companies engaged in the natural resources sector that would differ from practices in companies engaged in the manufacturing sector. This proposition motivated our selection of case companies representing four distinct industrial areas. A chemical company and the chemical business of a diverse multinational were selected to permit comparison of similar companies.

2. *Companies with greater consumer name recognition will have more protective environmental programs than companies with less name recognition.* Therefore, a consumer product company with a high degree of familiarity among the general public was included. Our interest was in determining the importance of protecting the company's good name in shaping EHS programs, and in contrasting the programs with those of other study companies whose consumer name recognition is low.

3. *The programs and decisions of small companies will be less protective of the environment than those of large companies.* A significantly smaller MNC in terms of employees, sales, and number of foreign facilities was included for the purpose of pursuing this proposition.

4. *Profitability of the corporation is important to strong EHS performance.* One case company is significantly less profitable than the other four, permitting examination of this proposition.

5. *Conformity with the EHS program developed by corporate headquarters diminishes as distance from headquarters increases and as cultural and political contexts become significantly different from those at headquarters.* This proposition motivated the basic protocol of examining facilities in both the United States and in case countries for each company. In addition, facilities in both developed and developing countries were examined.

6. *The greater the top management commitment, the better the EHS performance.* An effort was made to include companies with a range of top management engagement, in the expectation that this proposition could be tested.

7. *Having a well-publicized environmental incident in the corporate history is a strong catalyst to development of protective EHS programs.* Two companies with well-publicized incidents were included in the study in order to examine the impact of past problems on program development.

In Chapter 9 we will examine what drives a good EHS program. We use examples from the case companies, and rely on these propositions to frame the discussion.

In addition to the propositions, some exploratory topics are pursued. These are areas that we believe are worthy of examination; however, they are approached in an exploratory rather than an explanatory manner. Robert Yin offers an example to explain exploratory conditions:

> When Christopher Columbus went to Queen Isabella to ask for support for his exploration of the New World, he had to have some reasons for asking for three ships (why not one? why not five?), and he had some rationale for going westward (why not north? why not south?). He also had some criteria for recognizing the New World when he actually encountered it. In short, his exploration began with some rationale and direction, even if his initial assumptions might later have been proved wrong.[12]

Three main areas are the subjects of our exploratory inquiry. We had enough background and preliminary information to state a purpose and further define a direction, but because there is little research focused on these topics, we did not attempt to define a causal relationship and to put forth a proposition.

Management policies and structures. The organizational apparatus of an institution has an effect, at some level, on personnel and resource decisions. Do management structures in general impact EHS programs? Is the effectiveness of an EHS program different if it is managed in a decentralized versus a centralized structure? Policies articulate an agenda or a set of beliefs that can influence the behavior of those associated with it. How formalized does this articulation need to be to have an impact? What actions are necessary to accompany the written or spoken words to elicit a response among employees?

Standardization. The issues surrounding the development, implementation, and enforcement of global environmental standards, both internal to corporations and under international law, have been raised and debated. Are corporations developing uniform internal practices? Would uniform international standards benefit developing economies at the expense of the MNCs? Are there situations in which an inferior EHS standard is actually beneficial? What time frame and value criteria should be imposed when making these decisions?

EHS program components. A frequent assertion is that if an EHS program contains certain "key" components such as audits and line responsibility, improved performance will result. Which components or strategies are absolutely crucial to effective EHS programs? How do corporations define an effective program? What criteria should be used in determining consistency among these key components? How do corporations measure the effectiveness of their programs?

A final topic of interest does not fit the exploratory framework. The issue of *sustainable development* is explored in the cases; however, this inquiry is less directed than the previous ones. The discussion of sustainable development and its relationship to EHS management is an underlying and crosscutting theme throughout this research. We examine sustainable development not as a "stand-alone" condition but as a fundamental element or result of the propositions and exploratory areas presented so far and elaborated on in the chapters to follow.

SURVEY

The survey results discussed throughout this book are based on information from 98 companies that responded to a survey mailed out in the summer of 1990. Respondents were presented with 131 questions contained in a six-page survey.[13] The inferences that are made based on the survey analysis do not necessarily describe all U.S.-based multinational corporations; however, the responses do indicate interesting points.

Questions were included that examined such issues as how corporations are organized and staffed to address environment, health, and safety challenges; how environment, health, and safety policies are implemented in the field; and whether there are differences between U.S. and non-U.S. operations. Specific questions included: (1) what types of statements are contained in corporate EHS policies; (2) what factors prevent companies from doing a better job at EHS; and (3) how often environmental issues are addressed in decisions to acquire new businesses.

PROFILE OF SURVEY RESPONDENTS

The 98 companies responding to the survey ranged in size from 100 to 190,000 employees. The median number of employees was 2,800. The five groups shown in Figure 2.1 represent 20 percentile divisions of company size. Of the 98 U.S.-based companies, 96 percent had manufacturing operations in and/or outside of the United States. Only four companies had no manufacturing operations. One company had manufacturing operations only outside the United States. See Figure 2.2. Those with operations outside the United States had an average of nine facilities in an average of six different countries.

Key findings from the survey data will be discussed throughout the book to show how a large group of companies' policies and practices either support the findings of the five cases or contrast with them.

Figure 2.1
Company Size

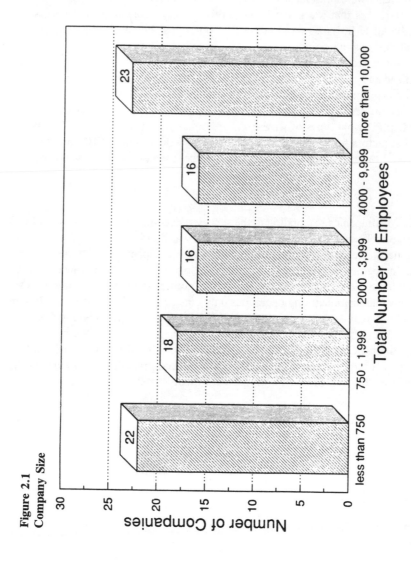

Figure 2.2
Location of Operations

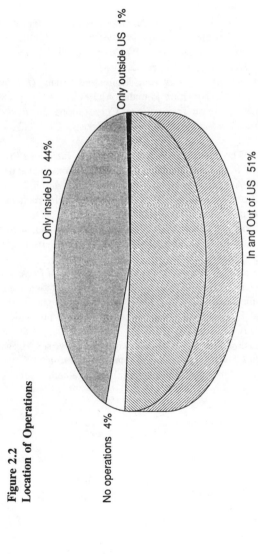

Only outside US 1%

Only inside US 44%

In and Out of US 51%

No operations 4%

25

NOTES

1. Dun's Marketing Service, *American Corporate Families,* vols. 1 and 2 (Parsippany, N.J.: Dun and Bradstreet Co., 1989).

2. Quantitative information on company operations may vary in the discussions that follow. This is because disparate reference years and sources are used in successive chapters.

3. Robert K. Yin, *Case Study Research: Design and Methods* (Beverly Hills: Sage Publishers, 1984), pp. 107–109.

4. Roger Kasperson et al., *Corporate Management of Health and Safety Hazards: A Comparison of Current Practice* (Boulder: Westview Press, 1988), pp. 119–132.

5. Thomas N. Gladwin, *Environment, Planning and the Multinational Corporation* (Greenwich, Conn.: JAI Press, 1977), pp. 234–242.

6. H. Jeffrey Leonard, *Are Environmental Regulations Driving U.S. Industry Overseas?* Washington, D.C.: The Conservation Foundation, 1984.

7. United Nations Centre on Transnational Corporations, *Environmental Aspects of the Activities of Transnational Corporations: A Survey,* ST/CTC/55, UN Pub. No. E.85.II.A.11, 1985.

8. Allen White, "The Transboundary Movement of Hazardous Products, Processes and Wastes from the U.S. to Third World Nations," Paper presented at the Annual Meeting of the Association of American Geographers, Baltimore, March 20, 1989.

9. International Labour Organisation, "Safety and Health Practices of Multinational Enterprises," (Geneva: ILO, 1984), p. 69.

10. Patricia S. Dillon and Kurt Fischer, "Environmental Management in Corporations: Methods and Motivations," (Medford, Mass.: Center for Environmental Management, Tufts University, 1991).

11. Mitchell Koza et al., *Company Policies for Environmental Protection: A Preliminary Study of Nine European Companies,* Report prepared for presentation at a meeting on Public Information: Companies' Organization to Deal with Environmental Issues, (*Paris, France*: UNEP, Industry and Environment Office, December 1989) pp. 4–5.

12. Yin, *Case Study Research*, p. 30.

13. For full discussion of survey design and methodology, see Margaret Flaherty and Ann Rappaport, *Multinational Corporations and the Environment: A Survey of Global Practices* (Medford, Mass.: Center for Environmental Management, Tufts University, 1991).

3

Policies and Structures

OVERVIEW

In this chapter we examine the development, implementation, and content of corporate EHS policies. We also describe the formal and informal organizational structures in companies for environment, health, and safety. The chapter concludes with a discussion of the challenges facing EHS professionals, looking in particular at the impact of reorganization trends on EHS management.

ENVIRONMENTAL POLICIES

There is significant interest among environmental professionals regarding the existence of EHS policies that have international implications. More specifically, there is interest in what the policies consist of, as articulated and promulgated at corporate headquarters, and how these policies are interpreted and carried out in both home and host country facilities.

One minimum test of top management commitment is the presence or absence of a written corporate EHS policy. Corporate policy statements can be an effective means for communicating the overall thrust of EHS intentions. If given prominence and operational components, these statements can have meaning to employees, other companies, customers, regulators, judges and juries, litigants, stockholders and stockholder groups, boards of directors, banks, insurers, community pressure groups, and the media.

In our survey, respondents were asked whether their companies had a written policy concerning EHS issues. Ninety-five percent of those responding said their companies had a written EHS policy in place. We also asked whether certain features were present in the EHS policy and had respondents indicate the importance of these features to EHS management. We have chosen three examples to illustrate some important contrasts. We asked whether the following appeared in companies' EHS policies: (1) an explicit statement to comply with existing U.S. laws; (2) an explicit statement to meet or exceed U.S. laws overseas when foreign laws are less stringent; and (3) a commitment to immediately halt operations if unacceptable EHS conditions are found.

Figure 3.1 shows a comparison of the percentage of respondents whose companies had a written policy containing the three characteristics with the percentage of respondents who felt that the characteristic was very important or essential. For example, the survey asked whether the company policy has an explicit statement to comply with existing U.S. law, and it also asked respondents to indicate how important they think the characteristic is to the management of EHS regardless of whether it is in the company policy.

Respondents attached equal importance to complying with U.S. laws and halting operations if unacceptable EHS conditions are found. Although fewer than half of the companies had a written statement about halting operations, more than 80 percent of respondents' companies had a statement about compliance. There is also a gap between the perceived importance of an explicit statement to meet or exceed U.S. laws overseas when foreign laws are less stringent and having a statement to that effect. We speculate that these results reflect the evolving nature of EHS programs. We believe that if the same survey were administered again in a few years, it would show that statements have been added to address policy areas where there is currently a discrepancy between perceived importance and the presence of a policy statement.

POLICY STATEMENTS OF CASE STUDY COMPANIES

Household Products had a written policy that its staff described as "not a flashy document" and that, until recently, was "largely unchanged over the past 25 years." Parts of the policy were rewritten by a vice president in 1983 and call for employees to "follow the letter and the intent of the law in every country in the world in which they operate." This policy has subsequently been updated and a new policy has been adopted that represents a "life-cycle" approach. The intent is to take into account environmental effects throughout an entire process: from product development, to manufacturing and distribution, through consumer use and disposal. This policy has been translated into 27 languages.

The policy of Oil and Gas does not explicitly refer to international aspects. Interviews and documents suggest that Oil and Gas environmental staff interpret the policy to mean that the company should adopt a posture internationally in which environment and health are protected by standards that are "functionally equivalent" to U.S. standards. The corporate staff claims that foreign management is asked to complete research on local requirements. When local requirements are compared to regulations for a representative facility in the United States, the differences are "usually found to be negligible."

Figure 3.1
Comparison of Written Policy versus Relative Importance of Issue

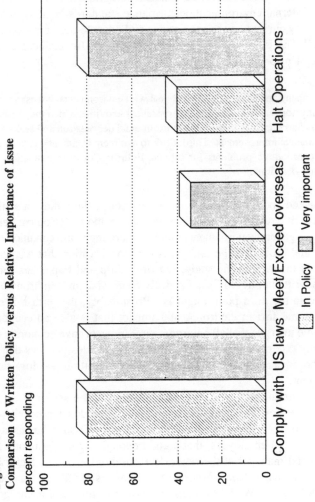

percent responding

In the Mexican facility of Oil and Gas, we found that the manager of EHS was familiar with the corporate policy. However, equivalency was not really an issue, because (1) the Mexican laws are largely based on U.S. Environmental Protection Agency (EPA) standards and (2) the facility goal of "zero discharge" is considerably tougher than either U.S. or Mexican law. The real questions in this case concern implementation and enforcement.

Although Pulp and Paper does not have an explicit international clause in its policy statement, it does include a separate paragraph that requires all employees to "adopt internal environment performance standards in situations not covered adequately by current law or regulation, and educate our employees on their application." The Pulp and Paper policy statement also includes a most unusual treatment of prioritization, as follows:

> [The company and its employees] will deal with environmental issues in the following order of priority: First: Take action to prevent activities which may result in health hazards. Second: Seek to prevent environmental degradation and seek out areas for environmental improvement. Third: Seek to avoid environmental actions which may produce economic problems for others. Fourth: Consider aesthetics in making operating decisions.

When asked about the inclusion of aesthetics in the policy, a senior Pulp and Paper executive said that it had been his idea to add it. "You can tell a lot about a company's general environment attitude by looking at the condition of the facility buildings and grounds," he said. This is an observation that we heard repeated many times throughout the study, but only Pulp and Paper has formalized the idea in its written policy. It is not entirely clear what an "environmental action" mentioned in the third point might be. Presumably, the term relates to actions that have some sort of environmental impact that is also an economic impact. It is possible to argue that all environmental impacts have economic impacts. No other policy statement of the study companies includes, as does that of Pulp and Paper, the idea that employees should seek out areas for environmental improvement.

When discussing policy matters in a meeting of the senior U.S. plant people at Pulp and Paper, the plant manager said: "We don't need elaborate policy statements like this from headquarters, if the CEO says he insists on ethical behavior from all employees, that's good enough for us." In contrast, both the environmental manager at the stateside Household Products facility and an environmental manager from corporate headquarters of Oil and Gas said, in exactly the same words, "We follow corporate policy as long as it makes good business sense."

Chemicals has recently adopted a policy statement that is comprehensive in the sense that it contains specific operational aspects and, in fact, is entitled "Policy and Operating Guide." It is of interest that the environment and the safety and health policy statements are separate, and that the environment policy statement, although quite full, contains no operational guidelines, whereas the safety and health policy statement is quite short but is accompanied by four pages of specific operational guidelines and includes responsibility assignments for the department itself, division presidents, line managers, and supervisors.

Chemicals has no specific reference to overseas operations in its policy. It does state that the company will "be in full compliance with all applicable regulations and, in unregulated situations, with [our company] guidelines."

Instruments does not have a written comprehensive EHS policy. The company does, however, have a written chemical safety policy that covers right-to-know issues and other aspects of toxic and hazardous substances in the workplace. The policy identifies responsibility for establishing procedures and sets forth the basic rules governing labelling, disposal, spill procedures, and other information.

ORGANIZATIONAL STRUCTURE

Organizing a business across national borders has never been a simple exercise. Before the recession of the early 1980s, many multinational firms adopted the so-called matrix system of management. The system is established by creating reporting lines for each middle or line manager that lead in two directions, sometimes more: toward a boss at the head office for each product or function and toward a chief executive of the local subsidiary who, in turn, reports to a head office, often through a regional headquarters. The intent of matrix management is to recognize that a given individual may perform more than one function and that for different functions, different reporting relationships are appropriate. Matrix management can, however, be confusing and overly bureaucratic. Companies are now modifying this approach in an attempt to simplify life and to take advantage of the new communications and data processing technologies in a much leaner and more competitive setting.

One idealized approach to organizational structure was offered by Kenichi Ohmae in a *Harvard Business Review* article. In describing the transition that companies must make in moving toward globalization, Ohmae notes:

> Decomposing the corporate center into several regional headquarters is fast becoming an essential part of almost every successful company's transition to global competitor status. . . . The more successful a company is at bringing both operational and strategic responsibility down to the regional or local level, the more likely it is that local or regional concerns, attitudes, affinities, and allegiances will shape the decisions of its far flung management cadre. . . .
>
> A global corporation today is fundamentally different from the colonial-style multinationals of the 1960s and 1970s. It serves its customers in all key markets with equal dedication. It does not shade things with one group to benefit another. It does not enter markets for the sole purpose of exploiting their profit potential. Its value system is universal, not dominated by home country dogma, and it applies everywhere.[1]

At present, most companies, including the five in-depth cases in this study, are not global and are still in the process of either restructuring or consolidating earlier moves. Consequently, the classic organizational division—between support and production personnel; between working, supervisory, and executive functions; and between headquarters, division, and facility functions—is the one within which most EHS managers must function.

COMMUNICATION CHALLENGES

MNCs are confronted with the need to operate facilities in remote settings where language and culture may be quite alien to headquarters managers. Companies are evolving means of communication that permit the headquarters managers to understand the operating situation, permit the plant managers to understand headquarters requirements, and permit all to understand and act when problems arise.

The language issue itself has never been an easy one. Setting and enforcing home country EHS standards in a culturally different host country situation can be even more difficult, especially because home country laws are debated and elaborated in a specific cultural, economic, and social context that may or may not be relevant to conditions around host country facilities. For example, such seemingly routine health and safety procedures as wearing respirators when performing certain tasks around volatile chemicals can present very difficult situations in countries where religion dictates that men wear beards. Furthermore, in developing countries, for instance, the concept of risk, especially cancer risk, may not be as meaningful to local communities as risks from job loss, from hunger, or even risks from loss of honor and status. During one interview at a facility in South America, a plant manager, when asked about installing a ventilation system over a solvent operation, replied, "You want me to worry about TLVs[2] when half my work force shows up each day with a belly full of worms!"

The language issue can also become quite literal. During an audit of a Brazilian facility, the U.S. EHS manager asked to see a list of the EHS program elements at a manufacturing operation. There was much confusion among the Brazilians as to exactly what he was looking for, although he repeatedly requested the "program elements." After discussion among the Americans and the Brazilians, a facility chemical engineer left the room and returned with a periodic chart of chemical elements, familiar to all of us from high school chemistry class; however, it was a far cry from what the U.S. manager had in mind.

CORPORATE CULTURE AND THE ENVIRONMENT

Often, the very large multinationals consist of a number of different businesses grown or acquired over time. The overall corporate culture can consist of a range of subcultures differentiated by these various businesses. At General Electric, for instance, the aircraft engine culture is quite different from the lighting culture; in Pulp and Paper, there are significant differences between the timber cutting group and the paper manufacturing group. Within Oil and Gas, there are important differences between the financial "deal makers," the petroleum prospectors, and the chemical manufacturers.

In terms of EHS programs the different subcultures can be important, because most often top corporate management is drawn from one of the various businesses. Although a neutral or vaguely negative attitude toward EHS matters at the board or top management level may not affect the day-to-day running of these programs, these attitudes can clearly influence the allocation of new resources to an EHS program and can have a major impact on the perceived status and prestige of the senior EHS executive. In addition, the implementation of corporate EHS policy within the division itself can be effective or ineffective, slow or fast, based on the traditional subcultures involved. All of this obviously occurs in spite of what the corporate hierarchy looks like on paper.

REPORTING RELATIONSHIPS

This section begins with an exploration of case companies' formal reporting relationships for environment, health, and safety and then examines the implications, with the recognition that many of these relationships are currently undergoing transition.

In Oil and Gas, the highest individual with EHS responsibility is a vice president, who in turn reports to the corporation's general counsel. The general counsel in this case is a member of the board of directors. In Household Products, there is a manager of environmental coordination, reporting to the chief executive officer (CEO) and serving a liaison function between line business areas and corporate staff. Chemicals, like Oil and Gas, has an overall corporate structure in which the corporate headquarters acts much like a holding company for a group of highly autonomous business units. The top EHS person in Chemicals is similarly located within the holding company corporate offices at the level of vice president. Chemicals, in the late 1980s, hired an individual to direct a newly created Office of Environmental Policy. This individual, who brings to the corporation a certain amount of name recognition, functions more in a public or political relations position rather than an operational one.

Pulp and Paper's top EHS executives are dispersed into regions, holding the titles of regional environmental managers. However, as noted in the discussion of policy, employees of the firm have been accustomed to hearing environmental concerns from the chief executive officer, and his leadership on the issue is a factor cited as important by those who were interviewed.

Instruments, a smaller firm than the others, has its corporate manager of environment, health, and safety reporting to the facilities manager, who in turn reports to the vice president for corporate communications.

Regardless of the specific location of corporate environmental functions on the organization chart, a critical factor in determining the effectiveness of environmental programs is the relationship between those on the corporate staff and those in line management positions.

LINE AND STAFF RELATIONSHIPS

Historically, tensions have existed in corporations between the staff and the line personnel, and the EHS area is an excellent case in point. The line people are the ones turning out product, making money for the company. Corporate EHS people are staff with responsibility for supporting the line function, yet with such relative intangibles as image, policy, and compliance; EHS programs may be perceived in businesses as having a negative impact on profitability. The distinctions between staff and line personnel can help explain why, without exception, all the EHS corporate managers we interviewed identify salesmanship as an especially valued skill for accomplishing their objectives. One corporate EHS manager observed: "EHS people need to have a strong technical background and management experience. You've got to have run a plant or had operational experience somewhere along the line. You have to be able to communicate effectively with the operational people."

Although greater autonomy and independence go a long way toward motivating managers and supervisors, it means that corporate EHS managers, with the need to measure and control and to limit risk and liability, have an especially sensitive and demanding job. As a result, many of them have developed a vocabulary to describe their relationship to division and facility managers. We have heard the relationship described as a "calibrating" function (Household Products); a "coaching" function (Pulp and Paper); an "orchestral conducting" function (Allied-Signal); a "counseling" function, and an "older brother" relationship (Oil and Gas).

HEADQUARTERS AND FACILITY RELATIONSHIPS

Line and staff demarcations in sophisticated operations may become increasingly blurred as management strategies change. However, there is also a real tension in corporate management between headquarters and facilities. It is at the facilities, after all, where all of the serious day-to-day production and value-added work goes forward. It is also at the facilities where organized labor has its input to the process. But it generally is at headquarters where the overall strategy for the corporation is developed.

Solutions to environmental problems may prove particularly challenging to the traditional headquarters/facility relationship. For example, pollution prevention activities require extensive process-specific knowledge to be effectively applied. Headquarters can encourage facilities to conduct audits and identify opportunities for pollution prevention; however, only facility personnel with an intimate knowledge of the production process, waste streams, and effluent discharges can effectively identify and prioritize measures needed to prevent pollution in that facility.

SAFETY AND HEALTH VERSUS ENVIRONMENTAL MANAGEMENT

Relative emphasis on environment, health, and safety varies across companies. Each of the functions has its own corporate history, having arisen out of a combination of union pressures, government regulation, competitive circumstances, and variations in the progressiveness of chief executive officers and boards of directors.

Within the United States, management and control of safety and health have been driven historically by the increasingly complex Occupational Safety and Health Administration (OSHA) regulations[3] and by union demands. Beyond these forces, however, productivity and, therefore, profit are enhanced by limiting accidents and sickness in the work force. Process safety, in particular, was institutionalized in corporations considerably before either occupational health or environmental management, with the gradual development of industry-wide good practice standards and separate professional codes, many of which have become international.

Because both occupational health and safety are regulated by the OSHA, it has made sense in the United States to manage these two functions in an increasingly integrated fashion within corporate headquarters. For one thing, contacts with regulators can be made more efficient and interdisciplinary communications within the corporation can be more straightforward.

The management of strictly environmental issues, on the other hand, has traditionally been the control of events "outside the factory walls" and has had a more recent history and an independent evolution. Because environmental management has been associated with the need to monitor and retrofit manufacturing processes, most environmental functions fell within the engineering disciplines at the facility level as recently as the 1970s. As regulations have grown in complexity and environmental issues are increasingly litigated in court and discussed in the media, both legal and public relations skills are brought into the environmental management equation.

In recent years, as corporations began to feel the need to respond more effectively to environmental issues, an environmental management function was seen to be useful at the corporate level, especially in companies with large and diverse manufacturing facilities to control.

Organizationally, the case companies, as well as others interviewed, are in different stages of integrating environmental policy and procedures with health and safety policy and procedures. Allied-Signal Corporation, for example, has roughly 400 professionally trained people who spend at least half their time working in an EHS discipline. The majority of these people are nurses, and the second largest number are safety personnel. There are very few industrial hygienists, and they are used "very sparingly" on the kinds of problems they are best suited to solve.

In the case of Oil and Gas, at the outset, the corporate environment function was relatively independent of the safety and health function and was headed by an engineering professional with process experience. The group has since matured to the point where it is headed by a vice president who is responsible for environment, health, and safety.

We were very interested to learn how companies made decisions on EHS staffing levels at their various operations. In the survey, respondents were asked to indicate the basis for determining facility EHS staffing levels by specifying the three most important considerations from a list of nine choices: risk of the operation; number of units produced; number of employees; country of operation; proximity to sensitive areas; legal and regulatory requirements; one (EHS staff person) per facility; profitability of facility; and cost of EHS staff.

As shown in Figure 3.2, the risk of the operation and the legal and regulatory requirements were most frequently chosen as the most important reasons for EHS staffing levels.

Survey data indicates a positive association between the size of the EHS staff and the amount of staff time spent on inspecting/auditing facilities outside the United States. Companies were divided into four groups based on the size of the EHS staff.

Among companies with operations outside the United States, those with smaller staffs were less likely to spend time inspecting/auditing facilities. This was a statistically significant difference.[4] Figure 3.3 shows that of the companies with facilities outside the United States with an EHS staff size of 20 or more, 82 percent responded that they frequently or occasionally spent time inspecting/auditing facilities. For those companies with facilities outside the United States and an EHS staff size of less than 3, only 42 percent responded that the staff frequently or occasionally spent time inspecting/auditing facilities.

Safety and health personnel have in common with environment managers the need to continually develop new arguments to support the expansion, or even maintain the status quo, of their operations. As an executive from General Electric told us:

> You need to communicate in language the businessman understands. In [one facility], for example, we have a predominately older work force, and the number of injuries in the muscular/skeletal area was very high. When we talked to management there about making changes, we did not focus on the workman's compensation issue (which clearly has financial implications), instead we talk about productivity and quality, which have an immediate impact on the thinking of the business people.

Our research in host countries indicates that overseas, safety policy and procedures receive greater resources and attention and are more advanced in terms of integration within operations than environmental policies and procedures. Quite often in developing host countries, we saw that fire protection and plant security were the dominant aspects of safety management. For example, the Mexico City environmental contact person at one case company referred to himself as a "fireman," and he attends frequent meetings in the United States on various aspects of fire protection.

Figure 3.2
Basis for Determining EHS Staffing Levels

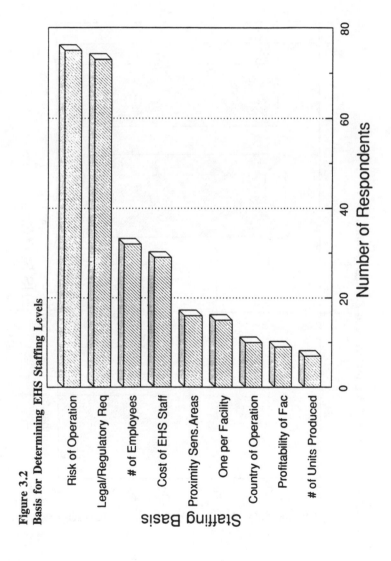

Figure 3.3
Staff Time Spent Auditing/Inspecting at Foreign Facilities

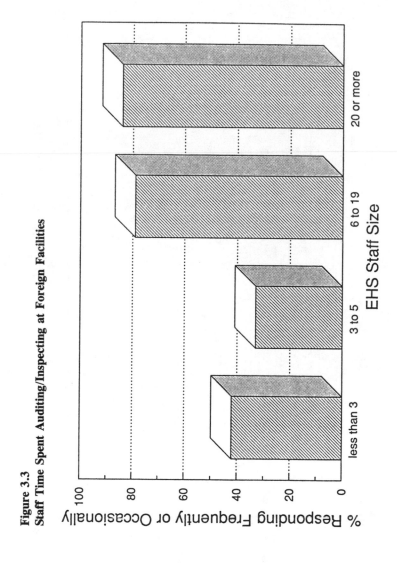

RESTRUCTURING OF EHS MANAGEMENT

As a result of the recession in the early 1980s, many MNCs undertook a basic review of their management structures, and a number of them went through a process of heavy payroll reductions (especially at the corporate level), integration of functions, and consolidation of lines of business.[5]

The leading MNCs, in an effort to regain the competitive position lost to the Japanese, looked to Japanese management styles and the new breed of management strategists, including W. Edwards Deming, and concluded that the "quality management" approach is an important and useful concept. This strategy calls for "collapsed hierarchies" and pushes responsibility as far down the line as possible, viewing each element in the production chain as the customer of the element preceding and striving for total quality in each step. It means giving "ownership" to supervisors and workers and as much autonomy as possible to facilities. It implies deep surgery to middle management functions and much greater breadth of oversight to the remaining corporate managers.[6]

Two study companies, Pulp and Paper and Household Products, have adopted the Deming approach formally and have put in place structural and functional changes to implement the theory. In 1990, Oil and Gas formally recognized the quality management strategy; however, the company would probably argue that even before taking this step, its very lean corporate management structure and the autonomy of its various businesses and facilities accomplished a similar end result without a structured Deming endorsement. Although Chemicals recently reorganized its EHS function, putting greater responsibilities out to the field, none of the managers interviewed referred to the Deming approach as such.

Another feature of this kind of management restructuring is a much greater reliance on both ad hoc and formal "teams" to address certain aspects of the management function. Pulp and Paper has a number of issue management teams, which are formed around issues such as the dioxin problem. Household Products has a less formal and still developing corporation-wide risk management team.

EHS PROFESSIONALS

As commonly defined, EHS professionals would include nurses; specialists in occupational health, process safety, product safety, industrial hygiene, and environment; chemists; and engineers from the relevant disciplines. However, when one considers the variety of tasks EHS personnel are called upon to undertake (for instance, in an inspection of a facility about to be purchased), a whole range of other skills and experience is relevant. In the United States, it is important to know the history of the site and whether or not hazardous materials are present in the surface or subsurface soil and water. Other potential liabilities, including those from radon gas and asbestos, must be identified, and tests must be correctly interpreted. In addition, facility environmental managers, as a result of SARA Title III, must now become skilled at press and community relations. Finally, as we noted previously, salesmanship has become an increasingly valuable commodity.

As companies restructure and decentralize and as regulatory complexity increases, the need for a management capability to coordinate the many professional functions becomes acute. In addition, at the corporate, division, and facility levels, senior and middle EHS managers need to interact frequently with finance, production, real estate, marketing, sales, and human resources departments. Externally, there are an increasing number of trade and ad hoc industry groups with which the EHS manager must function. Internationally, this requires well-developed interpersonal skills and sensitivities.

EHS managers, then, must employ many of the skills used by other managers in a complex organizational setting. In addition, they must address substantive issues of significant technical complexity within a framework of government regulation that is constantly evolving. The EHS policies and structures in place in a corporation are a reflection of the efforts that EHS managers have made to modify corporate culture and operations to more effectively protect environment, health, and safety.

The following chapter explores in detail some of the management challenges facing individuals in two of the case companies, Chemicals and Instruments.

NOTES

1. Kenichi Ohmae, "Planting for a Global Harvest," *Harvard Business Review* (67) 4 (1989): 136–145.

2. TLVs are threshold limit values, which are used to determine safe exposures to occupational chemicals.

3. OSHA was created by the Occupational Safety and Health Act of 1970 and currently has jurisdiction over 5 million workplaces where approximately 62 million workers are employed.

4. This was shown to be a statistically significant difference with $x^2 = 8.6$, df $= $ p $\leq$ .03.

5. Philip R.Harris, *Management in Transition: Transforming Managerial Practices and Organizational Strategies for a New Work Culture* (San Francisco: Jossey-Bass, 1985), pp. 1–29.

6. See, for example, W. Edwards Deming, *Out of the Crisis* (Cambridge, Mass.: MIT Center for Advanced Engineering Study, 1986).

4

Management Challenges in *Chemicals* and *Instruments*

These cases were prepared to stimulate discussion and are not intended to illustrate either effective or ineffective approaches to the situation presented. The material is based on interviews conducted in a real company; however, fictitious names have been used.

INTRODUCTION

The material presented in this chapter highlights the EHS challenges at two case companies, Chemicals and Instruments. The cases are based on company documents; library research; and extensive interviews with corporate officials, facility staff and management, and representatives from governments, nongovernment organizations, trade associations, and the academic communities.

The objective of the case studies is to present information that illustrates the complexity and multifaceted nature of the management of environment, health, and safety from various perspectives within a company. Our intent is to create a snapshot with enough depth and background to make the picture real.

The cases do not provide a comprehensive analysis of a specific company, nor is there an intent to make a judgement as to which corporations' environment, health, and safety programs are best. We have, however, used the information to illustrate some commonalities as well as distinctions in the manner in which EHS issues are managed in the study companies.

CASE STUDY OF *CHEMICALS*

After undergoing a hiring process that included interviews with every company president, and in some cases two interviews, Tom Blackwell felt that the process the company went through was an indicator of the importance Chemicals attaches to his responsibilities as director of environment, health, and safety. "They don't want me to fail," he observed. Since Blackwell took the job in 1984, significant expansion in EHS staff has occurred. Between 1986 and 1990, 54 full-time equivalent additions were made, which Blackwell characterizes as "indicative of the company's commitment: It costs a lot of money to add that much staff." He added that the company "made a conscious decision to improve in this area."

Blackwell is positioned to play a pivotal role in making progress in Chemicals' environment, health, and safety programs, since decisions on priorities, approaches, and managerial tools are primarily his. Blackwell's options include a wide range of program elements, and the challenge is to craft the strategy that will work best for Chemicals.

Company Background

Chemicals is a business within a larger corporation that acts primarily as a holding company. The holding company's activities are concentrated in four areas: a worldwide chemical business focusing on specialty and agricultural chemicals (the focus of this case study); energy-related natural resources; health care; and specialty businesses.[1] With facilities in 44 countries, total worldwide employment in 1988 was 45,700.[2]

In its Form 10-K filed with the Securities and Exchange Commission for the year ending December 31, 1988, the holding company reports being involved in lawsuits related to asbestos exposure from products previously sold by the company, and it acknowledges having been named by the U.S. Environmental Protection Agency as a "potentially responsible party" at over 20 sites where hazardous waste was disposed.[3]

The holding company is headquartered in New York, and Chemicals is headed by a president who is located at the same address. There are 13 divisions in Chemicals, each of which is headed by a president. Although Blackwell functions as corporate staff, his office is located in another northeast state, near five of the divisions. Blackwell observed: "There are a lot of turf battles that go on between Chemicals and corporate headquarters in New York. My boss is in New York, but the people who do EHS are here. My clients are the divisions." Blackwell's supervisor is Chemicals' vice president for environmental affairs. Blackwell added: "In EHS, we have more than enough to do. Congress is mandating job security. The system is on our side; we do not have to get involved in turf battles."

Culture

Chemicals is a highly decentralized company, and the divisions are very autonomous. According to Blackwell, "In this company, you have to be financially astute. Top management may not know about water pollution, but they know about every penny made in Canada by 10:00 in the morning."

Ron Eliot, an environmental lawyer for Chemicals, indicated that the decentralization resulted in significant differences among the divisions. He observed: "We are so diverse that you would not even recognize our South Carolina operation as being part of the same company." With this decentralized situation, Eliot said, "The challenge to us is to even out the performance across the company. Overall, we are way ahead of where we were a few years ago."

Decades ago, Blackwell said, one of the holding company's early acquisitions had the practice of obliterating the names on incoming chemicals and labelling them "vinegar" or "molasses" and also had a policy of prohibiting visitors in the plant. He offered this information to illustrate clearly the "tremendous philosophical changes" that the company has undergone. The company experienced rapid growth in the 1960s and 1970s, during which time Blackwell characterized the safety function as having been "dormant." During the 1970s, concern for environmental issues "increased, but was still small."

EHS Program

Important features of Chemicals' environment, health, and safety program include written policies, training, auditing, and a semi-annual environmental update meeting. Blackwell emphasized that facility notices of violations and fines are summarized and reported monthly and tracked for corrective action.

Chemicals has two general policies, one for environment, and one for safety and health. Both are dated 1988; however, they differ in level of detail. The health and safety document totals four pages and contains general policy and operational details, including information on responsibilities of particular departments, divisions, and individuals.[4] In contrast, the environmental policy is a one-page statement in which the company (1) recognizes and accepts its responsibilities as an environmental steward; (2) commits to full compliance with applicable regulations and, where feasible, will anticipate new regulations; and (3) supports environmental legislation and expects to participate in its development.[5]

The following sections focus on two managerial approaches currently in use in Chemicals: audits and the environmental update meeting.

Audits

The audits at facilities in the United States are scheduled by Blackwell's group and follow approximately an 18-month cycle. The audit is "tailored to the philosophical approach of Chemicals," according to Blackwell.

The audit team consists of at least two people from Blackwell's group. Depending on the nature and size of the operation being audited, as many as seven people may be involved in a team. They have taken people from other plants and have found that plant people are very enthusiastic about having the opportunity to see another operation. "No one objects to audits." Blackwell noted that in some cases they have plant managers asking them to come back.

Until very recently, after conducting an audit, Blackwell's group would write a report to the plant manager and his supervisor. They in turn would write a summary of findings, indicating what needed to be done and who would do it. The plant report was the only one that was circulated. This gave the plant the opportunity to rephrase or recast the audit findings. Blackwell observed, "We're not in the business of embarrassing people." Follow-up by Blackwell's group to determine whether the actions were carried out depended on where the plant was located. However, according to Blackwell, recent changes implemented during 1990 and 1991 require audit reports and compliance reports to be copied to division presidents routinely.

If major problems are identified as a result of the audit, they do not appear in the audit report; they are handled through other channels. Blackwell gave an example of one case in which there was no written report at all—he telephoned his concerns to his superior, and within a week of finding the problem, Blackwell had a meeting with the division president to resolve the issue.

According to Ron Eliot:

> We do not include a lawyer in our audits, and we do not use the confidentiality argument. This is my position and Tom agrees. If you are not willing to correct problems, you should not do an audit.

> We do not use the argument that audits are confidential for two reasons: (1) I do not think that the attorney-client argument will be successful if challenged, and (2) I think this would make my office a bottleneck for conducting audits.

> I have come to the view that I do not like over-lawyering. The environmental audit should be a normal part of your operations. To the degree that a lawyer is involved, it becomes abnormal.

Ron receives proposed revisions to the audit manual, and his comments generally reflect issues he identifies from his perspective of looking across the company's 13 divisions.

Semi-Annual Environmental Meetings

Twice each year Chemicals has a meeting in which all of the divisional environmental people are brought together to update one another and to discuss important issues. There have never been more than 36 people in attendance, and it is very informal. The table is arranged in a square so that there is no head, and everyone knows "that you're expected to tell what you've done for the last six months and what you plan to do for the next six," according to Blackwell.

The agenda for a recent meeting included a presentation by an outsider on risk communication, and a decision by the company to limit its waste contracts to two or three companies nationwide. Ron Eliot has often made presentations at these meetings, addressing legal issues that cut across divisions.

No one at that particular meeting was from the international divisions; however, there is usually someone there from South America, and Blackwell intends that the two European people will attend regularly. Blackwell said that this approach of having semi-annual meetings was in place before he arrived, and "it works. . . . If they decide to do something it gets done."

Management Information

Blackwell "receives no compliance information from divisions." Divisions that have "real issues" write monthly reports, but this information "won't filter up. We take the position that it should be kept at the plant level."

Blackwell had a book on his shelf by Frank Friedman called *Practical Guide to Environmental Management*.[6] Friedman is vice president of Occidental Petroleum, and the book outlines the system being implemented in Occidental: Data on environmental excursions at all locations is regularly entered in a centralized computer system and is available to corporate EHS staff. Blackwell said that they were "headed toward the Occidental system." However, with respect to the centralized receipt and oversight of compliance information, which is a key part of the Occidental system, Blackwell said: "I don't see that here. We have restrictive [water pollution control] permits at four of our facilities, and it would generate too much data. We have no reason to function the way Goodyear does, where all of their decisions are made in Akron. Here we are paying plant people to make decisions. If they're not, we should hire kids."

According to the attorney, Ron Eliot: "Other companies are a lot more formalized than we are. In any situation you might have to balance a general directive, corporate culture, and the need to write it down. You need to have some method of communication to figure out who told people what." He added, "To me the most effective means of managing information is sensitizing people throughout the company." Eliot observed that training sessions are an important way of achieving this type of sensitivity, and he added that after a session, "people don't hesitate to pick up the phone."

Environmental Goals

Blackwell said, "We are looking here for zero excursions. Inspections for RCRA[7] and OSHA[8] should result in only minor deficiencies. If there is a serious deficiency, pay or position would be affected."

Asked whether the goal was zero excursions at any price, Blackwell responded, "We mean that for a limited number of places." He cited three locations in the United States and all the facilities the company operates in Germany where "There is no upper limit for expenditures." In one of the facilities Blackwell identified in the United States, the plant has a permitted discharge to a river, and downstream is the intake for a city drinking water supply.

In terms of specific programs, Blackwell characterized himself as "very sensitive to underground storage tanks. I wrote a letter recommending action that has been taken in many places as policy. We have managed to get most of the tanks out of the ground and we are ahead of most people in that area." According to Blackwell, there were 217 underground storage tanks in Chemicals in January 1986, and in January 1991 there were 66 underground storage tanks.

Another program that has received a great deal of attention among environmentalists and others interested in improved environmental management is pollution prevention or waste reduction.[9] This approach involves a variety of measures including improved housekeeping, process modifications, raw material substitutions, and product changes. Blackwell said:

> We are in the process of developing a waste minimization program. It is a long-term approach, and it has to do with our goals for zero emissions; you cannot separate the two. In some product lines and with some chemicals, companies had better be planning for zero emissions.

> There is nothing new about waste minimization. Generally, as a process changes, you look for improvements, one of which is reducing the waste. What is new is the pressure for documentation, and that lags programs by as much as five years. You are hoodwinking the public if you lead them to believe that waste minimization is something new. We haven't done a very good job on documenting here, but we are going to do it.

According to the company, total emissions reported under SARA Title III are as follows: 1987—263,737,667 pounds; 1988—16,464,803 pounds; 1989—15,154,091 pounds.

Financing EHS

All the expenses for Blackwell's group "are fully allocated to the divisions." In some cases it is reflected as part of the division's overhead expenses, and in others the expense is "pushed down to the plant." One of Blackwell's goals is that EHS should become an overhead expense and not be allocated to the divisions. Currently, Blackwell's group tracks the hours expended for each division. Estimates for each year are based on the prior year's records. The legal department is paid in the same fashion. According to Blackwell, "This system is what they're used to. Operating units want to know what it will cost them."

Reporting Relationships and Staffing

The role of EHS varies depending on the division. For example, one division is headquartered in the same location as Blackwell. The division has an active program for underground storage tank replacement, and the division people bring their plans to Blackwell and ask for his comments on them, and Blackwell makes recommendations. "In most places this review would be done very formally, but that's not the case with this division. We're very close. In a sense we're supplementing their staff."

The relationship differs across the organization. "[A second division] doesn't like the direction we're trying to push them because it will cost them money." In a third division, "We audit, provide oversight and consultation, and represent the corporation." Blackwell noted that the third division "just hired a top-flight guy who is starting to manage very effectively." There is no direct relationship on paper between the new hire in the third division and Blackwell's office.

In clarifying the relationship between his corporate staff and the division EHS activities, Blackwell said: "Anytime we're doing something for a division that represents half a person, we get on them to hire. That's where the resources belong."

Despite significant staff increases cited earlier, the EHS coverage in some areas may still be thin. Blackwell noted that "only a few divisions have a dedicated safety person."

Program Effectiveness

Part of the challenge for Blackwell is operating in an industry in which there are giants, such as Du Pont, who have been hailed in recent years for their decision to stop manufacturing CFCs ahead of any schedule formally agreed upon by governments, and for making considerable efforts to reduce toxic air emissions and hazardous waste.

In rating Chemicals' performance, Blackwell described a hypothetical rating scheme in which "Dow and Du Pont are close to 10. We don't aspire to match either company's record. We want to be a 9.5. Right now I would put us between 5 and 6." He also observed, "We're not as proactive as we would like to be, but we are out of the starting blocks."

International

Blackwell gave an example of their operations in the Philippines, a situation in which the plant manager felt that the facility was located too close to sensitive populations. "They shut the facility down and moved it to another location in the same country. They wouldn't do that if they were strictly into profit maximizing."

"Our marching order is to treat off-shore facilities the same as domestic," according to Blackwell. "They are not equivalent, but we give that to them as a standard to achieve." Other companies have set goals of functional equivalence, that is, a situation in which procedures and equipment may vary across locations to accommodate a variety of local conditions, but the net effect to the environment will be the same in all operations. Deciding how best to achieve this marching order is a significant challenge.

There are two divisions of Chemicals headquartered in Europe, with an EHS person in each division. Both the person in Paris and the one in Switzerland have been with Chemicals for less than a year, and both travelled to Blackwell's office to meet people. He said, "They're still feeling their way around in terms of the exact relationship with this office. The mechanics are not in place yet." Blackwell observed of the European EHS managers: "People there are a little more demanding, they are paid more and they are better educated. They report to the division presidents."

Regarding audits outside the United States, Blackwell said, "When we started the audit program, my boss and I went to Europe and told them what we were doing and recommended that they do something similar." The company hired a consultant to conduct audits in Europe, and the new EHS managers will play an important role in deciding the next steps. Determining how Blackwell can effectively support the efforts of the new European EHS managers is part of his challenge.

The View from Switzerland

The recently hired EHS manager based in Switzerland, Richard Heiner, came to Chemicals from Dow Chemical, and he sees several differences between the companies. One specific example he offered was in the nature of the business; where Dow is a primary producer, Chemicals has many divisions that use purchased raw materials to formulate products. He observed: "Being a specialty chemical company, we do not have the problems of people making the basics. As a processor, we have fewer dangerous situations. The biggest emission source in one factory is the solvent from printing on plastic bags."

Management Controls

In describing how the environment, health, and safety program is being operated in his division, Heiner said:

> We have operations in 12 countries so I cannot check for compliance everywhere. In the German company I can, based on past experience operating in that country. The German facility happens to be our biggest as well. I am familiar with the requirements of the United Kingdom and the European Community, and beyond that, one can inquire about pretty superficial things. My boss is Dutch, and so he looks at our operations in Belgium and Holland. We try to have a variety of contacts or to use consulting firms to assist us in this.

However, Heiner indicated that he had encountered some problems in implementing the audits that Blackwell supports. "Not long ago a call was placed to our facility in the United Kingdom to find out when they might be able to schedule an audit. The response from the facility was 'I have had enough of audits and I don't want any more this year.'" Heiner added, "We do not yet have a routine schedule for environment, health, and safely audits."

According to Heiner, implementation problems are not limited to the facilities. He indicated that his division's middle management plays a role in frustrating changes that could improve the company's response to environment, health, and safety challenges.

> Two people from a consulting firm went around conducting audits. They were paid for by New York. They spent two to three days at each site, made recommendations and prioritized them. That's when the trouble starts. Managers need money to do these things, and they apply for funds and they don't get approvals from headquarters. I sometimes have the feeling that we all work for accountants here. Unless the authorities demand it, it won't happen. For example, in one case the consultant recommended n-hexane monitoring, this is not required by local authorities so the company will not make the expenditure. The audits have shown where we are weak, but this is just a snapshot.

Creating a situation in which divisions support expenditures that are not mandated by government yet will benefit environment, health, and safety is important to Heiner's strategy, and it is an area in which he indicated changes are occurring.

Financial Involvement

One approach to integrating environment, health, and safety into the decision-making process is reviewing major expenditures for their impact. According to Heiner, "We are the channel for all capital appropriation requests. There is a whole list of signatures required, and we sign off on anything over $50,000. We just started being in this process; our involvement is about two months old." This involvement has already had positive impacts in environment, health and safety. Heiner said:

> For example, there was a request for replacement of a machine that was too loud, so I discussed various alternatives with the person making the proposal. The production person making the request was concerned that the total cost of the replacement would be 5 percent greater if a housing was built around the machine or if it was relocated so that the impact on people was reduced.

Heiner identified other changes that are also emerging. "Originally, if a capital appropriation request was for health, safety, or environment, there was no obligation to calculate the payback. So I asked, 'Why not?' I wanted to look at these expenditures as an investment as well." This approach may not always show that investments for environment, health, and safety meet the general criteria for payback used in other company decisions; however, Heiner offered one example in which the environmental investment was a good one.

A proposal came through to deal with the emissions from a printing press. One option proposed capturing and burning the fumes, and there was another destruction option. I said, "Why not examine the option of solvent recovery?" They found that the payback has been superior and they are making money on solvents.

Company Culture

Developing an environment, health, and safety program in a decentralized corporation is challenging. Heiner observed:

> At Chemicals, you must have individual approaches, and programs cannot be directed by a central man. You need local approaches. If I were in [some other company] my job would be totally different; I would say, "do this, this and this" and it would be done. Here, I must try to convince everyone that EHS is part of their job.

He added: "I am getting resistance, so it must be working. I always say the worst they can do is fire me, but I have never been fired from a job yet."

The academic training of top decision makers may be a contributing factor in explaining why financial concerns seem to dominate environmental issues, according to Heiner. He observed:

> No one in [top management] in Chemicals has technical training.[10] There is a very strong financial orientation. In European American companies if you can make the thing work, you are promoted. In a European company, you have to have the technical degree, you have to be Herr Doctor.

Heiner offered some examples of his own to reinforce the point Blackwell made about attention in the company to fiscal matters:

> Some time ago I wanted to attend an international conference on occupational health to renew old and make new contacts and to have two days to concentrate on the same issue. In Dow, there would have been no question of my attending something like this. Here, only the vice president can sign for such an expenditure. He made such a noise about spending 500 Swiss francs that I finally decided not to attend. "You have to learn to live with Chemicals culture," the vice president said.

> I also have to get approval for the purchase of books. I submitted a request to buy the current version of the Merck index, and I was questioned about the contents of the book and why I needed it. When I indicated that it was to replace the outdated volume that I had, I was asked why the older version would not do. Junior people at sites can order books themselves.

At the same time he cited examples of the day-to-day frustrations with Chemicals culture, Heiner indicated his clear understanding of the larger issues with respect to corporate environmental practice.

The issue of competitiveness is raised by our people all of the time—they feel they may be at a disadvantage relative to those who spend less on the environment. Bhopals happen because of this attitude. The German chemical industries have signed an agreement not to produce anything anywhere in the world with lower environmental standards than are in place in Germany. You *can* do it—it costs a little more money.

However, Heiner also observed: "Attitude change can be accomplished without spending a single cent."

The View from Paris

Heiner's counterpart in Chemicals' other European division is Jean Morel, who also recently joined the company. Morel, who has worked in a European-based multinational chemical company and has done consulting work, often talks with Heiner as they seek ways to integrate environment, health, and safety considerations into the company's decision-making process.

Jean Morel pointed to figures from the 1988 annual report indicating that the European operations under the joint responsibility of Heiner and himself contribute half or more of the revenues for their divisions of Chemicals. He added:

Chemicals has two European divisions with about 7,000 employees. We represent about 33 percent of the profits of the company as a whole. We therefore have an important impact on profits, but most of the focus of audits and such is on the stateside divisions, not in Europe.

Asked whether he attributed this situation to differences in regulations, Morel responded: "No, it's not a multinational view. As long as the profits keep coming in, they just keep the operation going. There is no motivation to change."

Company Culture

Describing the culture in the company, Morel said: "Fundamental changes are needed at Chemicals. I think that the culture has to evolve to take into consideration EHS. Business plans should consider EHS." He added: "There is enormous resistance to change." Morel indicated that, from his point of view, specific changes are needed from the top:

The difficulty starts in New York, where health, safety, and the environment are not in the business plan, and promotions are not influenced by the EHS record. If New York sorts itself out, we have a chance of getting somewhere. Some people are responsible and know that the times are changing . . . [but] many of the old timers are resistant.

Morel continued by describing his efforts to achieve greater management control over environment, health, and safety considerations by changing division procedures.

Management Controls

Morel's division has 15 manufacturing sites, and his responsibilities include product safety, worker health and safety, environment, fire protection and loss control (insurance issues), and security.

> I have been trying to get every site to have a follow-up system after accidents. My first meeting was on that subject. No, I do not go back and make sure that the appropriate changes are made to prevent similar accidents in the future. I have to leave that to the sites. Unlike Heiner, I do not feel I can be directly involved in this. When I visit the sites, I check to see that the system for conducting follow-up is in place.

> I ask that the incident report be used for *anything*, whether it is a spill, injury, fire, or release. We are now trying to do two reports, one for failures, and the other for follow-ups. It is difficult to get a response, so I now count it as a plus if a report is filed.

In environment, health, and safety, Morel said:

> We are gradually making people expand their reporting systems. I changed the report from one page so it is now four pages in order to add, among other things, details on causes behind injuries. The president indicated that he was concerned about the increased level of complexity, but it is considerably less complex than the financial reporting. If you want to achieve the same level of information, you would have to multiply by 20 the amount of EHS information to make it comparable to the financial information.

In general, Morel is following a strategy of using systems in place in the company and making modifications to address EHS. He observed: "Because of the resource constraints, I look at the management aspects of the problem." Offering an example, he said: "My big problem is action plans. The company uses action plans for quality management, so I use the same approach in safety."

Implementing the audit program poses challenges to Morel as well as his colleague in Switzerland, Richard Heiner. The same consultant that performed audits in Heiner's division also was retained by corporate headquarters in New York to perform audits in the facilities for which Morel has EHS responsibility. According to Morel, "The plants did not ask for the audits so they just filed them away and did nothing with them. Now I am bothering them about it and I am becoming unpopular with a few people—I take that as a sign that the bite is starting to hurt."

As Morel continues to develop EHS programs in his division, systems for audits, like systems for following up on accidents, will be developed.

Annecy, France—Plant Visit

In the past, Morel said, the operation in Annecy had lost money for many years, and indeed it may still be losing money, but Chemicals has been interested in keeping it going because as a matter of business strategy it wanted to have a position in France. According to Morel, "The company has a philosophy of not throwing good money after bad, so although they kept the company, they put little money into it. The people there have been ingenious in making do with the best they could. I have heard from many sources that it has improved a lot."

The Annecy plant makes products for the automobile industry, such as sealants against noise and vibration, anti-corrosion compounds, bonding and sealing compounds (for example, the material that holds the windows to the metal in a car), coatings that contain polyvinyl chlorides, and anti-grit materials that help keep small rocks from chipping the paint off the lower portions of the car body. In addition to auto products, the plant produces a variety of sealers for the construction industry (to paint on wet basement walls to keep out moisture, for example), adhesives, and epoxies.

Some of the products go out in bulk to automobile manufacturers and other large users; however, other products are put in retail-size packages right at the plant. Some of Annecy's customers are the European counterpart of an individual Discount Auto Parts or a Grossmans, so in one warehouse, these orders are assembled and shrink-wrapped onto pallets or put in other containers for delivery.

There are 170 people at the site and they run two shifts. The site is in a developed area with both residential and commercial neighbors. The area is not entirely fenced; however, near the entrance are prominent signs that prohibit smoking.

The manufacturing performed at this site consists of taking a variety of raw materials from many chemical producers and mixing them in machines of different sizes and types, depending on the type and amount of product that is desired. There were ventilators over the tops of many of the mixers that were intended to function as hoods, but the tops of the machines themselves were open in many cases. There were several powders or dusts spilled on the floor around the mixers, and few of the workers were wearing protective clothing; some may have been wearing safety shoes, but not all were wearing gloves or aprons.

The plant manager said that masks or respirators are worn when big bags are poured (no pouring operations were observed), but he added that it was impossible to have workers wear masks all of the time because their jobs require a great deal of lifting and bending, and are physically demanding, especially during the summer when it is hot.

Workers use large wooden scrapers to remove material the consistency of bread dough from a huge mixer—like a scaled-up equivalent of a KitchenAid mixer. After the scraping is done, the machine is wiped inside with cloth or paper and then washed down with solvent prior to starting a new batch. Several of the workers had the compounds caked liberally on their jeans and tee shirts.

Despite Blackwell's interest in underground storage tanks, those at Annecy, aged 20 to 25 years, have not been tested. The plant manager observed that it was impossible to test a tank in place for integrity; however, Morel indicated that he was familiar with in situ testing technology that is in use in the United States. The plant manager indicated that they planned to do something about the tanks within the next two years.

In one location, a pneumatic machine is used to put epoxy product in small packages for consumer use. It was difficult to talk in the area because of the machine noise, but workers were not wearing hearing protection. The plant manager said that they had measured the noise levels in the work area and found that levels at the part of the machine where product and packaging are fed in were high enough to require protection; however, levels right where the workers stood most of the time did not. "We found that protection was not necessary, and so it is not used," he said.

Emergency Procedures and Staffing

According to the plant manager, not all of the workers on the site had received training on procedures to follow in case of a major accident or fire. He added that "This is a problem because right now there are a lot of temporary workers on the site, and they have not been trained."

The plant manager added that it was very difficult to attract workers to the plant when they could commute elsewhere and make a much higher salary. A new employee parking area was being built at the front of the facility, and the plant manager expressed hope that it would make working there more attractive.

CASE STUDY OF *INSTRUMENTS*

Frank Perroy thinks about his job title and reporting relationships often. He knows that the top environment, health, and safety person is a vice president in some corporations. Frank has the top environmental job in Instruments and responsibility for the technical side of health and safety, yet he does not even report directly to a vice president. He was eager to help Instruments' new CEO see the value of his position, but he was uncertain how to make the strongest case: by pointing out the sound business thinking behind some recent environment decisions, or by articulating the advantages of elevating EHS to a higher level.

Frank grew up with Instruments. He started in the research group soon after he completed his chemical engineering degree, and he expects to retire from the company. For 10 years he worked his way up within the organization. He became a manufacturing supervisor, where many of his responsibilities involved ensuring the company was in compliance with air and water discharge permits.

The quantity and complexity of Instruments' products and processes paralleled the growth of the state and federal environmental regulations. Frank became so adept at the regulatory and enforcement components of the company's manufacturing that he actually worked himself out of one job and into another. His responsibilities as a chemical engineer in charge of manufacturing were overshadowed by Instruments' need to develop a more comprehensive environmental management program. Much of the impetus for the creation of his current position, manager, environmental services and energy conservation, is credited to Frank's advocacy, and a bit to the "default" principle. In effect, he wrote his own job description eight years ago.

"Theoretically I'm in charge of the environmental operations worldwide," thought Frank, "but I've never even been outside the United States." Like so many issues, mused Frank, the EHS profession and responsibilities have "taken on a life of their own" even before the organization had grown enough to formally cope, let alone manage.

Frank's troubles were actually more systemic than simply not having seen the foreign facilities' EHS practices. His location in the corporate hierarchy isolated him from many long-term planning decisions and minimized the reach of his authority. The status quo with regard to environmental issues at Instruments had been very low-key. The management motto was in effect: "If you don't want to read it in the paper, don't do it!" His title and job responsibilities, combined with the increasing visibility and liability of environmental concerns, made it critical that he elevate the status and decision-making influence of his position.

Company Background

Instruments is a relatively small multinational corporation, with 6,650 employees worldwide.[11] The company designs, manufactures, and markets instrumentation systems and other related products and services for the management and control of industrial processes. Founded just after the turn of the century, the company has been publicly owned since 1958. It owns plants in six countries and has approximately 100 sales and service centers worldwide.[12]

Twenty years ago when Frank started with Instruments, most of the process controls the company manufactured were pneumatic, electromechanical, or mechanical. This required milling operations, painting facilities, plating lines, and sheet metal operations to build housings for the equipment. Process controls now consist primarily of a variety of increasingly sophisticated measuring devices, linked by silicon technology to an overall system for control; computer hardware and software play a critical role.

Once a system is set up for a client, changes can be made by replacing printed circuit boards and/or reprogramming software. The majority of the manufacturing is in the United States, and none of it is as labor or as capital intensive as it used to be. In addition, the company now performs far fewer "dirty" processes while manufacturing the controls. For example, many of the metal housings have been replaced with rigid plastic units supplied by outside vendors, thus decreasing the need for metal painting operations that require extensive pollution controls.

The company has a range of over 1,000 products used primarily by the chemical and petrochemical, pulp and paper, mining and metals, food, and electric utilities industries. Products are also used in textile, cement, rubber, and glass processing; water and sewage systems; other systems related to the control of air and water pollution; scientific laboratories; and ambient air measurements.

A breakdown of the Instruments market shows that of the 1989 orders, 47 percent were from the United States, 22 percent from Europe, and the remaining 31 percent from the rest of the world. In the same year, the chemical industry made up approximately one-third of the company's orders, with the oil and gas and pulp and paper industries sharing a third.[13]

Frank recalls some shaky financial times for what he thinks of as a family-oriented, "Yankee values" company. During the 1970s much of Instruments' work was focused on serving the energy-related industries. "We're really led by the nose by those oil and gas industries," was what management claimed. The energy crisis and changes in the nuclear industry decreased demand for the type of products the company provided. When the company attempted in the early to mid-1980s to return to a more diversified, smaller contract-oriented business approach, these markets had been filled by their competitors and by the emergence of new competition from computer companies. Instruments was not alone in this situation. According to an industry publication, "Competition is continuing to rise steadily for the process control system vendors. The instrument companies, who long dominated the process control market, originally with their analog controllers and later with their distributed control systems, are seeing some of their market slip away."[14]

In reflecting on some of the more turbulent economic times, Frank remembers the EHS programs being characterized by upper management as "a protected species." Furthermore, he was told that "safety is an employee issue and very fundamental to our business. Environmental issues are a 'no options, no choice, no cutback' matter."

Recent Managerial Trends

Attempting to stabilize the financial foundation of the company, management has recently undertaken substantial restructuring programs, the most recent giving a boost to the new service-oriented approach. This was accomplished by reorienting overseas facilities from manufacturing to engineering and service facilities. It required providing these facilities with training, installation, sales, service, and research and development (R&D) bases. During a one-year period, the company spent over $50 million to downsize facilities and reduce personnel.[15] In addition to these programs, the company has embarked on a series of cost containment programs, resulting in a 35 percent reduction in the work force between 1981 and 1989. The company is now a more service-oriented business with less emphasis on basic manufacturing.

As Frank was taking note of changes in the financial status and the product orientation of the company, he also considered how the recently appointed CEO was developing a new organizational structure. This included moving toward more matrix management, with an emphasis on cross-functional activities, smaller operating staff, quality circles, more rigid reporting requirements, and more delegation of responsibilities. More employee input was being built into the system through the formation of Quality Improvement Process teams. Sixty-six teams were in place or being started up by the end of 1989, and the goal was to develop another 74 teams in 1990.[16]

Frank's thinly staffed office consists of, in addition to himself, an administrative assistant and a technical assistant; therefore, most of the information he uses he gathers himself. This is fairly consistent with Frank's "management by walking around" style. However, the overseas facilities are managed and run by local nationals and there are no designated EHS professionals at the foreign operations. Frank's information is funneled down to him by upper management at corporate headquarters. He characterizes the top managers as a group of well-rounded professionals whose backgrounds include research, finance, production, and manufacturing. Most of these eight to ten management positions are occupied by people who have been with Instruments for 20 to 30 years.

Instruments/Beijing

Frank picked up the majority of his information about the Beijing plant over the past few years from Brad Parks. Brad had returned to corporate headquarters after a three-year stint in Beijing to establish a joint venture with 51 percent ownership by the Peoples' Republic of China and 49 percent by Instruments. The remainder of Frank's perspective on Instruments/Beijing came from a one-month visit to the main Instruments facility by a contingent of 50 Chinese employed at the Beijing plant. The thought of these Chinese front line managers and supervisors brought back memories. For example, piling the visitors into a van and taking them for a traditional New England clam bake, and raising eyebrows in town when it became known that they had brought along their own cook from China.

Then, Frank remembered, there were the training courses given to these managers at the corporate classrooms. There was an interpreter present, and the instructors were told to keep their sentences short and concise, but Frank thought of the sessions as somewhat stressful. Although it was challenging, Frank helped the group go through the key points in an emergency response manual. It was not clear that the specifics of risk procedure translated all that well. Frank gave the Chinese trainees copies of the Chemical Safety Guide; however, it was in English.

The Instruments facility in China raises many issues for the corporate decision makers. The company feels pressure from the various countries in which it operates and sells, to do the manufacturing in those locations. "Developing countries want factories," Parks said, and he added: "No country will allow you to participate in the marketplace without some value-added." Although the term *manufacturing* has many definitions and variations, Frank sees Instruments as defining it to the advantage of the company in particular situations, and actually he would characterize the bulk of its overseas operations as "assembly." Further, thought Frank, with the new corporate realignment and restructuring, the emphasis on a service-oriented approach means that the offshore operations will reduce manufacturing activities and concentrate on application, engineering, and service functions.

This approach raises further issues and Frank recalls a discussion with Brad. Brad has argued to the Chinese officials that Instruments "wants to transfer technology at a higher level" by providing meaningful employment to skilled engineers, rather than hiring a factory full of workers to run a plating operation. Brad feels the benefit to China is greater with the Instruments approach because when and if the company leaves, it leaves behind a pool of people with "reusable intelligence," and he characterized the new Instruments approach as one in which the company is "willing to provide greater benefit to the country at less risk." In addition, Frank feels the approach would result in less impact on the developing country's environment and fewer adverse health effects than might be expected from a more traditional manufacturing situation.

One specific example of the new approach being used by Instruments was a decision about electroplating at the Beijing facility. Metal plated parts were going to be required by the Beijing operation, and Instruments made plans to have the plating done on site. Waste from a plating line consists primarily of water contaminated with various metals, acids, bases, and oils; treatment processes are available to remove metals, balance the pH, and thereby greatly reduce the toxicity of the waste. The Chinese government set effluent limitations for wastes of this type that were stringent, but within the capability of available technology. Instruments felt it had to make the investment in a good process for the value-added and to meet the guidelines for effluent set by the government.

Frank Perroy was asked to design the treatment plant, which he did, and he was ready to go to China to supervise its installation. At the last minute, the company decided not to construct the treatment facility and to purchase plated parts from a local supplier in Beijing instead. The total cost of building the system Frank designed was $150,000, and the current annual cost of using the outside shop is $20,000.

Although he was disappointed not to go to China, Frank could see the rationale behind the decision. Granted, you are trading off hands-on control over the quality and the schedule; and granted, the payback for the treatment system he designed was shorter than some investments the company had made in environment, thought Frank. However, he thought, "We are in a high-tech business, and products are changing quickly. Sometimes we have to look for a one- or two-year payback."

On the matter of increasingly relying on suppliers to provide parts, Frank had concerns about how suppliers manage environment, health, and safety. Brad made it clear that "we don't want it cheap if it's cheap because the vendor is dumping waste in his backyard." Parks characterized this as a pragmatic business decision rather than an ideological position: A vendor who is not complying with laws could be shut down by the government. Instruments would then have to go elsewhere for parts, perhaps losing time and jeopardizing delivery commitments.

Another challenge at the facility in Beijing has less to do with business strategy and more to do with the workers themselves. He recalled Brad Parks's description of what he found when he first walked into the Beijing plant: "It was hot, and most of the workers were barefoot, not even wearing sandals. They were cooling themselves with a series of small electric motors to which people had added fan blades fashioned by hand." According to Brad, the Chinese managers recognized that there were safety concerns and in effect said to him: "We have a problem here; how do you handle this in the United States?" So, Instruments tried to "put together some safety standards that we thought were practical for the environment in which we were operating." The Chinese "see a need, and there is a desire there to fix it," Brad said. He characterized this as a situation in which local managers "use Americans for leverage."

It took Instruments two months to secure shoes for the workers at the Beijing facility, and Brad made the decision to pay. He said later: "As part of its social welfare program, the company bought the shoes." Brad did not involve his boss because he did not see any other way to deal with the situation. Corporate culture at Instruments emphasizes quality of the environment for individuals, and he characterized this decision as "not debatable."

Safety glasses were another area in which Brad had a struggle. They were made available to workers, and their use required, but compliance was not high. According to Parks, "I'd walk into the factory unannounced once or twice a week to check, and if it was really bad, I'd just shut off the main power until everyone that was supposed to have on glasses, put them on!" He added: "I also talked with the production manager about the situation, emphasizing that it was her responsibility both to act as a role model and to ensure that workers were in compliance."

Frank was familiar enough with the situation, and he knew that "even next door in our [Instruments/Home] facility we still have people who neglect to wear basic eye protection." It was something he addressed on his frequent walks through the plant; however, the issue of protective equipment might benefit from attention at a higher level, he reflected.

As Frank continued to piece together his thoughts on Instruments/Beijing, he jumped ahead to reviewing what he knew about the London facility.

Instruments/London

Frank felt somewhat more in touch with the activities at the London plant for a number of reasons. The age and size of the facilities and the manufacturing operations of the Instruments/London and the main Instruments facility that Frank walks through daily are very similar. Also, he has had more frequent and in-depth conversations with management concerning EHS issues at this location. Although Frank regrets the limitations on his travel, most of the top corporate managers travel to the London site annually. The feeling of having a more accurate picture of the EHS activities was a situation of false security, however. As he pondered what he really knew, in terms of daily activities, Frank became aware of the fact that it was fairly minimal. Possibly, he reasoned, this familiarity has led to an unwarranted comfort level.

Philip Noble, vice president at the London operation, said at one point: "Sure, we run a good, safe place and meet all the national laws—I mean, certainly the intent of the law, maybe not to the letter." Noble also said: "Maybe if we closed down the London site the air would be cleaner around the plant, but the social environment would go bananas! We believe in absorbing the local culture, not imposing the Instruments culture on the locals."

A challenge facing Noble is anticipating how Britain will respond to the environmental aspects of Common Market initiatives, and he observed: "These Common Market standards are providing targets that the United Kingdom and other European Community members can't realistically meet."

As the relationship between business and government changes in European Community member countries, perhaps Frank could serve as a significant resource to people like Philip Noble. Frank reflected that participating in development of regulatory policy and building a firm basis for anticipating where government is going next is an area he emphasized, and it is one in which he has acquired considerable skill over the years. A prime example is a decision the company made at a facility located near corporate headquarters.

Instruments/Home

Frank began to think about the $3 million the corporation spent to improve the treatment of its process effluent through hook-up to the municipal waste water treatment facility.

For years, Instruments/Home had been treating its industrial waste and discharging the treated effluent to a nearby lake in compliance with a National Pollutant Discharge Elimination System (NPDES)[17] permit. A while ago, a group of citizens became concerned about the quality of the water, and the company investigated the matter and determined that pollution sources other than Instruments' effluent had a greater influence on lake quality (runoff, septic tanks, etc.).

The citizen group hired a lawyer and Frank recalls a few tense meetings; however, "after a while they left the lawyer at home because they felt they could work with Instruments' management." Instruments had decided not to do battle on this one. "Let's face it," Frank said to his supervisor, "I can spend time in the factory or I can spend time with citizens groups—I belong in the factory working with manufacturing people looking for better ways to do things."

Despite the company's relatively small contribution to the lake's problem, a decision was made to commit funds to hook the company up to the municipal waste water treatment plant. Part of the rationale was that when the NPDES permit came up for renewal, the older treatment plant would eventually have to be upgraded to meet increasingly stringent effluent standards. Frank recalls management stating that the main reason for going beyond the minimum standards was to "avoid having limits placed on our production."

Frank sees this expenditure as anticipatory from two perspectives. From the point of view of addressing environmental problems, Frank has told management that "if you just meet the regulations today, you'll be in trouble down the line. You have to get ahead, do more, and establish a margin of safety, and it will pay in the long run." The expenditure for upgrading facilities at corporate headquarters in the United States is also consistent with the corporate movement toward pulling out of an overseas reliance on manufacturing operations and "focusing efforts in our own backyard so we can keep an eye on things."

NOTES

1. *Moody's Industrial Manual* (New York: Moody's Investors Service, 1988); Holding Company Annual Report, p. 1.

2. Holding Company Annual Report, 1988, p. 1.

3. Holding Company Form 10-K, 1988, pp. 13–15.

4. "Safety and Health," Policy and Operating Guide, Chemicals, January 1988.

5. "Environmental Policy," Policy and Operating Guide, Chemicals, January 1988.

6. Frank B. Friedman, *Practical Guide to Environmental Management* (Washington, D.C.: Environmental Law Institute, 1988).

7. The Resource Conservation and Recovery Act (RCRA), first passed by Congress in 1976 and amended subsequently, is the basic federal law regulating disposal of solid and hazardous waste.

8. The Occupational Safety and Health Administration (OSHA), was created by the Occupational Safety and Health Act of 1970 and currently has jurisdiction over 5 million workplaces where approximately 62 million workers are employed.

9. See U.S. Congress, Office of Technology Assessment, "Serious Reduction of Hazardous Waste: For Pollution Prevention and Industrial Efficiency," OTA-ITE-317 (Washington, D.C.: U.S. Government Printing Office, 1986).

10. Since the time of this interview, hiring has occurred at corporate headquarters to include an individual with a technical degree.

11. Annual Report, Instruments, 1989, p. i.

12. Ibid.

13. Ibid., p. 5.

14. "Market Outlook '89," *Chilton's I&CS,* January 1989.

15. Instruments' Form 10-K filed with the Securities and Exchange Commission for the fiscal year ending December 31, 1989, p. 2.

16. Annual Report, Instruments, p. i.

17. National Pollutant Discharge Elimination System is a permitting program established by the Clean Water Act. Under this program, liquid wastes can be discharged to rivers, streams, and surface waters if the content of specified pollutants is kept within parameters established by the permit issuing agency (usually a state environmental agency or the U.S. Environmental Protection Agency).

5

Environment, Health, and Safety Program Components

OVERVIEW

In response to the range of regulatory and other pressures, MNCs are in the process of reorganizing environment, health, and safety management capabilities. Some companies have developed fairly detailed programs that are similar in their various components[1] to those of other companies. These include the programs promoted in a number of public forums by companies such as Allied-Signal, 3M, Union Carbide, and Du Pont. Some of the commonalities of EHS programs can be attributed to the fact that companies borrow ideas from each other and utilize common consultants, and that, in the United States, regulations are sufficiently detailed that many program components owe their presence to compliance considerations.

Some managers take the view that program components by themselves are a reasonable guide to overall EHS effectiveness. However, we view components as a necessary but not sufficient indicator of potential effectiveness. In short, two companies with very nearly the same management systems can be quite different in terms of EHS effectiveness. Other factors—such as corporate culture, corporate management structure and function, home and host government regulatory enforcement, economics and culture, media and community pressure, and customer and stockholder attention—can and do have strong influences.

PROGRAM COMPONENTS

The emphasis in this chapter is on the pieces or components, the sum and composition of which are defined as an EHS program. The exercise of quantifying program effectiveness is primarily reserved for Chapter 9. EHS effectiveness has a number of meanings. A program that achieves compliance at the lowest cost is quite different from one that achieves compliance at a reasonable cost but also has a better probability of protecting the assets of the company. Both of these, in turn, are different from a program that achieves compliance at a reasonable cost, protects corporate assets, and at the same time reaches back into production and design to minimize waste, strive to achieve a goal of zero discharge, or purposefully search out opportunities to develop "green" products and markets. However, all these types of EHS actions can be in place and still work against sustainable development if large differences in local needs and perspectives are ignored, especially in developing countries.

A list of the corporate EHS program components most often specified by individuals in the case study companies, by other companies, and in the literature is provided in the following list.

High-level management commitment

Line management EHS cooperation

Incentive systems for performance

Formal internal reporting and disclosure systems

An independent EHS auditing program

Inspection procedures

Professionalism

Training programs

Internal EHS standards

Risk assessment methods

Measurements of EHS costs and benefits

A system to value EHS goals in business terms

A crisis management procedure

To some extent, the more components within an EHS program, the better the program. For example, training and crisis management planning have the potential to be more effective than crisis management planning alone. Also, some program components are much more important than others.

PROGRAM COMPONENTS OF CASE STUDY COMPANIES

The various program components and management strategies used by case study companies are developed to assure compliance and, in a few companies, to go beyond basic compliance and achieve a "proactive" posture, at least from a public relations perspective. One division environment manager indicated that considerable effort was required to achieve compliance. He noted that a significant portion of his time was consumed with regulations: being aware of them in detail and conducting efforts directed at being able to "just make compliance." He added:

"We have to be terribly efficient with our manpower, given all the paperwork and regulatory issues. Throw in new thinking and profit motivations and dinosaur industries, and there is no way we can assure compliance. . . . No violations is a good goal.''

Companies attempt a variety of management techniques and strategies, often accompanied by operational or systematic tools designed to carry out the intentions of management. The following discussion highlights EHS program components that have reached a level of familiarity with many company EHS personnel, with examples and references to the case study companies in particular.

Establish High-Level Management Commitment

At the outset of the study, this component was cited most often as a crucially important key to EHS management success. In his book on environmental management, Frank Friedman argues that commitment of top management has been vital in the development and implementation of an EHS program at Occidental Petroleum.[2] This view has been reinforced by the consulting firm Arthur D. Little,[3] and by Roger Kasperson et al., who note that "commitment of high-level management to health and safety appears to be an important, if somewhat elusive, contributor to effective hazard management. . . . Health and safety issues must penetrate the highest level of corporate decision making if hazards are to receive primary consideration.''[4] Friedman and others assert that a central indicator of this commitment is an environment committee of the board of directors.

Commitment at the top can frequently mean that financial and human resources will be directed to EHS programs and that many of the other components mentioned previously can be put into place. Of even more importance, perhaps, is that in instances of inside competition for scarce resources, EHS matters can get a fair hearing. Combined with a clear purpose, top management commitment can allow a total program that goes beyond straightforward compliance.

Among the study companies only Oil and Gas had such a top-level environment committee. EHS interviewees at Pulp and Paper, on the other hand, claimed that everyone in the company knew that the executive officer was deeply committed to ethical practice, and this included EHS activities. At Household Products, an environmental coordinator is formally charged as liaison between line business areas and corporate staff, reporting directly to the CEO.

Establish Line Management EHS Cooperation

Line managers are the ones through whom top management conducts the central production business in a manufacturing corporation. As staff, EHS managers by themselves are relatively powerless, even with top management commitment. The EHS manager of a major division in one company was referred to by one of the corporate lawyers as "the custodian with a tie.'' Without exception, our interviewees believed that cooperation from line management is essential.

As has been noted, EHS people are more likely to see themselves as facilitators, coaches, technical resources, calibrators, and salesmen. Among the line managers, the facility managers are probably the most important. In the domestic facility of Pulp and Paper, the plant manager said: "I am personally responsible for environmental matters at this plant. We have had a slow start, but now I'm starting to see some progress."

In the current atmosphere of downsizing, collapsed hierarchies, networking, and team management, with responsibility being pushed further and further out to the field, line management accountability takes on even more importance. However, increasing confusion and some stress are arising in a number of companies as new emphasis is being placed on individual responsibilities and leadership.

Establish Incentive Systems for EHS Performance

One test of the formality of a line manager's responsibility is whether his or her compensation is related to effective EHS performance. Among our study companies this was not uniformly the case. The rule seemed to be that compensation was based on overall performance, but EHS excellence was not specified. A usual situation is one in which managers are penalized for a history of compliance irregularities but are not rewarded for proactive behavior or programs.

One case company indicated that when its staff was thinking about tying bonuses to EHS performance, the prospect drew immediate attention: An industrial group leader, in reviewing a raise request for a facility head, asked what that facility's safety record had been. The facility managers were asking questions within the week. "You better believe that information got around fast."

Establish Formal Internal Reporting and Disclosure Systems

These systems have been central to management functioning since the advent of the modern corporation. With potentially expensive compliance violations, corporate liability and insurance requirements, and the need to measure and control expenses, reporting and disclosure systems can be crucial in EHS management.

One approach to internal reporting and disclosure has been adopted and circulated outside the company by Allied-Signal. This approach has internal disclosure at its center. In a program called "bubble up," all significant problems are supposed to rise to the appropriate level for action. Allied-Signal's system also relies on a "letter of assurance" program in which the president of each business unit receives a letter once a year from the plant manager or the operations head affirming that all of his or her people know what the corporate environment policy is, that training sessions have been held, and that the plant is in compliance with Allied-Signal policy and pertinent laws. The company claims that should problems be found, the letter includes a plan for addressing them.[5]

None of the study companies had such a detailed and formalized system of internal reporting and disclosure, and Allied-Signal itself acknowledges that such a system is highly dependent on corporate history and culture. Allied also says, however, that it believes its approach could be made to work in decentralized as well as centralized corporate situations.

All study companies are in the process of experimenting with various reporting systems that attempt to summarize operating and EHS results and trigger action according to some criteria. It is important to recognize here that in devising internal reporting and disclosure mechanisms, all companies, including the study companies, are facing the need to balance control with the new autonomy of facilities that comes with decentralization and downsizing.

Establish an Independent EHS Auditing Program

Among all the management and compliance tools that were discussed with the study companies, the audit was most on the minds of both headquarters and facilities personnel, especially in overseas facilities. The new autonomy of facilities, the significantly different operating environments in overseas situations, and "the heightened appreciation of U.S. law which transfers the tort liability of independent subsidiaries to the corporate parent if there is evidence of command and control by the parent"[6] have given rise to an almost universal movement among the larger companies toward EHS audits.

Audits can have a number of forms. Basically, headquarters EHS personnel, sometimes with independent consultants, sometimes with lawyers, and sometimes with managers from outside divisions, conduct on-site visits with a specific agenda of inspection and oversight in mind. Some are detailed compliance audits in which each reporting form, compliance record, and all physical operating situations are reviewed. Others are more management-focused audits in which controls and supervisory structures are covered. Ideally, results are brought to the attention of top management, and a program for correction is undertaken with follow-up schedules.

Headquarters personnel claim that facility managers welcome audits because they identify problems that plant personnel habitually overlook. Furthermore, headquarters EHS people claim that they can stimulate interest in audits with plant managers because if the need is found for management or equipment upgrades, support and dollars come from headquarters auditors. Although this seemed to be a factor in some facilities we visited, performance ratings of programs and personnel, especially in overseas locations, seemed to be the principal incentive for facility action.

All study companies have put audit programs in place, although some companies' versions are very informal and do not occur consistently with regard to intervals or locations. At Pulp and Paper, a specialized function at headquarters level has been established and specific individuals are charged with oversight. At the Oil and Gas facility in Mexico, plant personnel were preparing for an audit of a new industrial hygiene program. At the foreign facility of Instruments, EHS management was unfamiliar with the concept and practice of auditing. Household Products conducts regular audits of specialized functions such as process safety control, in addition to a broader environmental audit.

Several issues in the area of auditing remain controversial. Although the World Commission on Environment and Development's report, *Our Common Future*, and a variety of interested agencies and institutions have called for MNCs to make audit results public, there is very strong resistance to this among some study companies, especially among the corporate lawyers. The main fear is that if a problem is found and made public and then not corrected, serious liability will accrue to the owners and managers. It has been argued that because of the potential impact on communities, refusal to disclose audit results is an act against the goals of sustainable development. This remains a central and controversial EHS issue.

There is also disagreement over who should pay for audits, and case companies have differing views on this matter. On the one hand, some companies feel that if a plant is forced to pay for audits in some sort of "chargeback" arrangement, there will be little incentive for plant managers to seek out the corporate technical or management assistance that could be forthcoming. On the other hand, others contend that if a plant is forced to pay for the audit, it will value it more highly and take greater advantage of the process. One manager said that because it is all company money anyway, "Why play with funny money?" Audits in this respect are really a special case of how headquarters and facilities view one another. As headquarters EHS people are increasingly seen as providers of services and less as authority figures, the method of paying for these services becomes an interesting and potentially important question. According to the manager of one case company,

> A consulting firm did environment audits at facilities throughout the Division. They were even paid for by corporate headquarters. Never was there a follow-up on the audit by corporate [headquarters]. Only about half of the sites have a plan to execute the recommendations in the reports, and in fact, no one has told the facilities they have to follow up. Most of the recommendations are very straightforward because it was so easy to find recommendations to make. The plants did not ask for the audits, so they just filed them away and did nothing with them.

Establish an Inspection Program

More than any of the companies in the study, Allied-Signal (not a case company) singles this component out as an especially overlooked program component, using the motto, "The uninspected invariably deteriorates." Allied distinguishes inspections from audits. Among other things, inspections can be undertaken on a surprise basis. While some observers may consider this step to be somewhat prosaic, experience in the overseas plants of at least two of the study companies suggests that this is a potentially important and cost-effective program element.

In one plant of a case company, drums of unidentified chemicals were stacked in a vulnerable site, a shed roof was in bad repair after a rain, unused materials were randomly stacked in dangerous places, and various personnel were slouching in chairs reading magazines. In a developing country context where neatness is not a top priority this may not be surprising, but one would expect that such a plant would be a candidate for a serious EHS episode. Inspections could clearly make a difference here, combined with other crucial program components.

Establish EHS Professionalism

With the rapid disappearance of rigid hierarchies and the concomitant increase in teamwork, networks, and individual responsibility, loyalty to a unit, a plant, a division, or even an entire company is less and less likely to be a motivating factor in individual behavior. In these circumstances, adherence to a professional code of ethics and standards becomes a crucial management element. One EHS manager in Mexico had been to a number of fire safety training courses in the United States. He proudly referred to himself as a "fireman." This kind of professional pride can go a long way in advancing overall EHS goals.

Professional standards are also a reliable anchor when ambiguities exist between country and government expectations, or between plant and headquarters values. It is interesting to note that the National Association of Environmental Professionals in the United States has an accreditation system, and more and more corporate managers are becoming members.

Establish Education and Training Programs

Education and training applies not only to EHS personnel but to middle management as well. A USX environmental manager interviewed for this study noted: "We have a pretty good handle on good practice, but we have to keep providing new material for middle managers, otherwise their other responsibilities command their total attention."

Education can take several forms. One device involves the daily supervisor meetings observed in facilities in two of the study companies. In these meetings, status reports on various EHS matters were discussed together with operating and production reports, and each was given equal time and attention. The educational and motivational advantages of this procedure are clear.

The need to train plant EHS people to deal with the community and the media is being met by the use of outside consulting firms that instruct managers in the fine points of getting a story across to the media. These and other risk communication techniques are being introduced in a number of case companies, most notably Household Products and Chemicals.

Establish Internal EHS Standards

Government standards vary not only from country to country but also from one subnational jurisdiction to another; from state to state in the United States and from province to province in Canada. These differences can wreak havoc with compliance goals and cause expense and confusion. The problem has been overcome, in theory, by companies that have established their own internal standards. If these standards are set as high as those in the most stringent country in which the company operates, several difficulties can be overcome. As noted, Oil and Gas has a policy that calls for the use of equivalencies across national borders. They claim that in actuality the differences between countries are quite small, as many developing countries draw on U.S. and other Western regulatory models. A number of interviewees felt that it is only a matter of time before virtually all countries have roughly the same EHS standards; they believe that a proactive approach is to anticipate this in their own internal standards.

An open question remains as to whether some of the money spent by companies in developing countries to achieve the same high protection (for example, against cancer) that exists in the United States might not be better spent on community training and education and other infrastructure development. As of this writing, there is a drive for uniform global EHS standards among MNCs in countries belonging to OECD. If such standards were enforced evenly throughout the world, presumably more overall protection and greater economic efficiency would result.

Establish Risk Assessment Methods

Techniques and methodologies are available for assessing EHS risks throughout operations cycles and in project planning. Most troublesome are low probability/high consequence events. One case company had an event that resulted in more than 150 deaths. The assistant director for risk engineering commented that even with the event,

> people still think that catastrophic things cannot happen to them. We need to spend time on the break point between the cost and effectiveness of doing analyses and taking preventative measures. . . . People do not have a lot of time for "what ifs" and thinking about the low frequency events. It's not an attitude problem; you have to have an expert shake them up and wake them up.

In developing countries, corporate risk assessment is complicated by migrations of people from rural areas to urban centers. Former industrial zones are often surrounded by makeshift settlements, putting significant numbers of people in jeopardy. Whenever possible, risk assessment should take this phenomenon into account.

Establish Measurements of EHS Costs and Benefits

Environment, health, and safety is a particularly difficult area in which to evaluate costs versus benefits. A starting point among many MNCs has been to weigh the monetary value of assets protected and liabilities avoided against program cost. However, this can be a complicated business. As one division EHS manager stated the problem:

> "One of the things that you certainly can measure is preventable losses in the form of waste through accidents. . . . When you don't put a value on life, you will not make investments that are needed to protect it. In the present system in this company, the value of life is zero. You have to use nonprofit and loss investments for EHS expenditures in this situation, which is simply wrong for fire protection. They installed one protection system just because I recommended it, not because the financial implications were examined.

Establish Means to Value EHS Goals in Business Terms

This component is really a corollary to the previous one. Line managers will "buy in" to sometimes expensive and onerous EHS programs if they can see their business value. General Electric, for example, has instituted an in-house newsletter for distribution to the business sectors within the company. The first issue contained an article illustrating this strategy:

> [I]ndustry protested when California voters passed Proposition 65, an extremely stringent toxics control and consumer product labeling regulation. GE Plastics recognized that their polycarbonate water bottle business potentially could evaporate overnight because the plastic used to mold the bottles contained traces of methylene chloride, a suspected carcinogen regulated under the new regulation.
>
> Even more ominous was the fact that state initiatives sometimes propagate across the country and the entire food packaging business nationwide was potentially at stake.
>
> GE Plastics did not sit idly by and complain. They responded with process changes and reformulations that eliminated the targeted component chemical and not only maintained the product line but captured the market in California![7]

The idea of creating new products and markets driven by government regulations is not new, but the present research has not uncovered an equivalent example as a corporate EHS selling strategy.

Establish a Crisis Management Procedure

One aspect of crisis management is the type of emergency response planning and training that is required in the United States for facilities that handle hazardous materials and wastes. The case studies indicate that in this aspect of corporate EHS programs there is considerable variability from home country to host country.

For example, in a facility located in France, the plant manager said that employee training on what to do in the event of a major accident or fire was a problem because "right now there are a lot of temporary workers on the site, and they have not been trained." In Brazil a facility manager explained that the local population "just wasn't ready to know about risks." He felt that telling them would cause hysteria.

A joint industry-UNEP project, called APELL,[8] provides guidelines for emergency response measures and has been circulated widely to industry and government. This document and the accompanying programs have high visibility among the government regulators in Brazil and Mexico with whom we spoke. Oil and Gas helped organize and run an APELL workshop and drill in Brazil. They were further involved in sponsoring a workshop dealing with hazardous materials management that brought experts (fire chiefs) from the United States to Brazil. The MNC representatives as well as the bureaucrats frequently cited lack of infrastructure to support this type of crisis management plan.

SURVEY RESPONSE TO PROGRAM COMPONENTS

A set of survey questions investigated the frequency with which specific program components or tasks were addressed by the respondents' staffs. We were interested in learning not only how much time people spent on certain environment-related activities but also how the time allocation varied in other regions of the world where the company had operations. The questions were presented in the following two parts: (1) how often these tasks occupy staff time in dealing with the EHS issues of U.S. facilities, and similarly, (2) how often the same tasks occupy staff time when they deal with foreign facilities.

The tasks included: monitoring compliance with laws and regulations; providing technical consultation; soliciting input on the development of EHS policy; reviewing requests for capital expenditure; consulting on the design of education and training programs; inspecting/auditing operations; community and press relations; trade groups; and negotiating with government regarding EHS issues. The four response categories were: frequently; occasionally; rarely; and never.

Figure 5.1 illustrates the responses from the 50 companies having operations both inside and outside the United States indicating their staffs frequently or occasionally spent time addressing these issues.

Upon further analysis, some statistically significant[9] variation was found regarding the frequency of time spent addressing specific tasks at foreign operations, depending on whether or not the company had a vice president with exclusive EHS responsibilities. Figure 5.2 shows the responses of companies with operations outside the United States. In companies with a vice president of EHS, the staffs were more apt to frequently or occasionally spend time monitoring compliance, reviewing capital expenditure requests, and negotiating with governments in relation to foreign facilities than those with foreign facilities that did not have a vice president of EHS.

Figure 5.1
Comparison of U.S.-Foreign Staff Time Spent on EHS-Related Matters

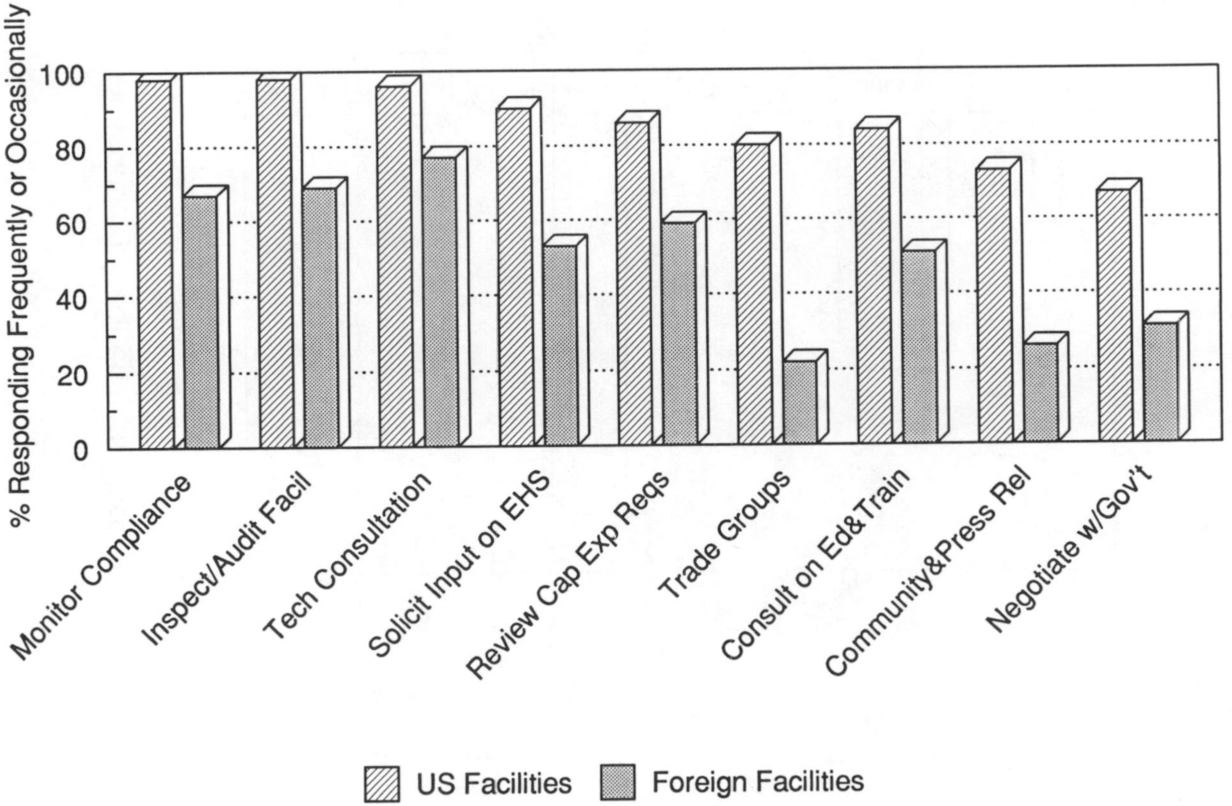

□ US Facilities   ▨ Foreign Facilities

Figure 5.2
Effect of Vice President of EHS on Staff Time Spent on EHS Issues Relating to Foreign Facilities

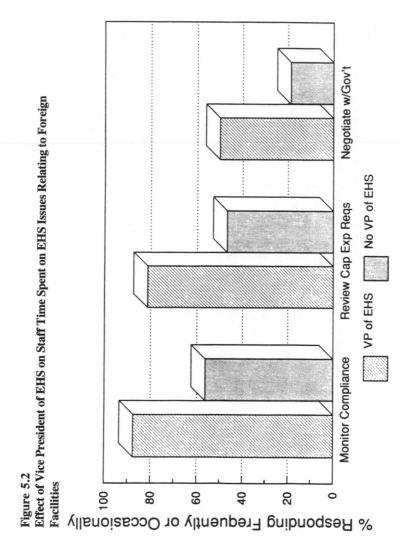

GLOBAL EHS MANAGEMENT PRINCIPLES

This short trip through actual company experience in applying various EHS program components, and analysis of the survey results, permit a revision and modification of thinking. It seems appropriate to break down a revised list of EHS components into the following subgroups: (1) basic program components—the specific on-the-ground components, such as training, that are directly under the control of EHS managers; (2) management strategies—more general approaches to successful implementation of the basic program components, some directly involving EHS managers only, but others more broadly based in the company; and (3) corporate or industry-wide goals—to advance EHS specifically and sustainable development generally.

The list is not exhaustive; many companies, governments, and others are in the process of extending and refining ideas. This is an iterative process that will no doubt continue for some time. With recognition of these limitations, the following discussion sets forth a range of components, strategies, and goals that potentially could represent a statement of global EHS management principles and practices.

BASIC PROGRAM COMPONENTS

Establish and Maintain a System for Compliance with Laws and Regulations

Such a system should recognize differences between countries, including language and culture issues. It should also recognize the need for constructive and supportive relationships with governments and communities. Increasingly autonomous facilities can be important factors in regional economies and cultures and can make significant contributions to local community life.

Establish Formal Internal Reporting and Disclosure Systems

These systems need not be elaborate but should provide managers with assurance that company policy and government regulations are being honored not only in fact but in spirit. Reporting by exception can be efficient, but a tracking procedure needs to be in place to cover prolonged nonreporting from facilities. A culture can be fostered that encourages internal disclosure, but it needs to be accompanied by a formal system that enables senior management to monitor compliance status.

Develop an Independent EHS Auditing Program

As authority and responsibility are increasingly being pushed out to divisions and facilities, significant pressure is put on operating people to act independently to meet company production and profit goals and at the same time achieve relatively complex company and government EHS goals. Reporting and disclosure systems alone are probably not adequate in these circumstances to assure the board of directors of sufficient protection from liability and compliance problems. A number of models exist. Most of them utilize not only lawyers but also operating people from other divisions. A punitive approach usually is not called for, but follow-up procedures should be built in. When plant managers see operating advantages to audits, they will often "buy into" the process.

Establish a High-Visibility Inspection Program

Physical inspections of facilities can be an efficient way to safeguard the assets of the company. The emphasis should be on improving practice and corrective actions, not on taking punitive actions. The inspection program should be accompanied by an action plan or maintenance program that has regular input as a central component. In many situations it would enhance the inspection program to extend the inspections to third-party waste transportation and disposal facilities. Varying the scheduling of inspections, as well as the individuals participating, would minimize the routine nature of an inspection program.

Give High Priority to Education, Training, and Retraining

Although this component is important in domestic facilities and for middle managers in line positions, it is crucial in overseas facilities where language and cultural issues can dominate. As a simple example, training materials should be in the language of the facility. If training materials are generated in-house, the possibility of marketing this information on a commercial basis might be explored.

Establish Internal EHS Standards

Experience strongly suggests that variations in government regulations and in enforcement can create ambiguities in compliance efforts in plants in different countries. Management efficiencies and control can be greatly aided by the establishment of internal standards. As previously noted, however, drawbacks include cost and risk of focusing attention on areas other than those needed for sustainable development, which can at times be counterproductive to the best interests of the host country.

Incorporate Risk Assessment in Operations and Project Planning

A risk assessment process that incorporates a risk management component can provide a rigorous scientific basis for specific EHS actions and expenditures. Experience in the case studies strongly suggests that in developing countries, the approaches used in OECD countries will need to be flexibly adapted to recognize divergent health and economic conditions and cultural values.

Plan for Crisis Management

The research strongly suggests that crisis management planning can be an important EHS tool if it is carefully tailored to local situations. In the United States and Europe, where a strong emergency safety and health infrastructure is already in place, evacuation drills and other crisis techniques can be effectively implemented. In facilities in developing countries, companies should probably carefully explore how they might be of use in helping to develop the country's infrastructure.

EHS MANAGEMENT STRATEGIES

Establish Formal and High-Level Management Commitment

Such commitment can take many forms. Written policies signed by the chief executive officer, the formation of an EHS committee of the board of directors, and affording high prestige to EHS management personnel are most commonly mentioned. The tests of commitment, however, are in the operating area, where the necessary resources are made available, where EHS managers are supported in "close call" judgement situations, and where deeply ingrained production and operations mind-sets are challenged.

Give Line Managers Formal EHS Responsibility

Responsibility for environment, health, and safety has been basically a staff function. This case study research strongly suggests that when line managers are given specific EHS responsibilities and are aided and supported logistically and technically by EHS staff, overall corporate policy is most likely to be advanced. In facilities in developing countries, where basic operating conditions are difficult at best and where production itself is sometimes a difficult goal, EHS objectives, which can often seem remote and inappropriate, have little chance of being realized without the full support of the plant manager and the line people above him or her in the division.

Develop, Encourage, and Honor Professionalism

This strategy has a number of virtues. Its efficacy in the health and safety areas, especially in developing countries, was noted in a number of the cases. Professional standards and recognition can exist independently of government or company pressures; they often have international status and are a source of prestige and pride. Professionalism within a company or within an industry sector can create a network to bolster high standards and facilitate flows of technical and management know-how.

Develop a System to Value EHS Goals in Business Terms

EHS managers have a better chance of selling their perspective and services to the business people in the company if they can develop a vocabulary and set of measures that are persuasive to this group. These can go beyond the asset and liability protection model to arguments showing competitive, marketing, efficiency, and quality advantages. If it is systematically and carefully exploited, such a strategy can offset, at least in part, the negative image that accompanies bearers of bad tidings (paperwork, fines, penalties, and compliance problems of all kinds).

Include EHS Performance in Incentive Systems

Bonuses, promotions, raises, and honors of various kinds continue to be the basic management incentives. Even where decision-making autonomy and pride of product are increasingly being promoted, managers who are tangibly rewarded for solid EHS performance will be powerfully motivated. These reward mechanisms should be applied not only for problems avoided (number of accident-free days) but also for innovation in process configurations, new product ideas, and the like. They can be especially valuable incentives in developing countries, where status and the symbols of status continue to be potentially powerful motivators.

Develop an EHS Management System Keyed to Corporate Goals

In the more progressive companies we interviewed, a number of managers advocated the development of an EHS management system, especially a behavior-based system. One advantage of such a system is that it can be monitored and fine-tuned *as a system*, not simply as a series of parts. When possible, such a system should be keyed to other overall corporate goals. For example, a number of MNCs recently began to develop variations on a total quality system for EHS, using teams and networks and moving away from more traditional hierarchical structures.

Experience suggests that facility personnel are sometimes slow to accept management innovations from headquarters. Corporate EHS managers may be in a position to be go-betweens in this process, if they can demonstrate clear advantages to plant managers and others. On the other hand, the cases show that the facilities can be more innovative and ahead of corporate thinking. Here, the go-between function might conceivably be applied in the reverse direction.

The study cases reveal instances in which innovation originated at one facility in a large, diversified organization and, through the good offices of corporate EHS, was broadened and incorporated throughout the EHS program of the entire company. New MNC downsizing and pushing out responsibility will put an increasingly large premium on the ability to seize this kind of opportunity.

Use Outside Management Consultants Opportunistically

In one case, consultants brought in to assist a company to downsize were used to convince top management of the value of the corporate EHS function, and most of the EHS programs were salvaged. This time-honored use of consultants to advance the in-house program of one corporate faction may be particularly timely, as many companies are currently in the process of downsizing.

CORPORATE OR INDUSTRY-WIDE GOALS

While certain EHS principles and practices, such as compliance, and a number of the other components and strategies listed here, are clearly in the hands of EHS staff, a number of crucial overall goals are not. This is particularly true of corporate actions that could contribute to sustainable development. Some of these are listed in the following discussion.

Select New Projects or Acquisitions Based on Sustainable Development Criteria

Often, EHS managers are brought in to evaluate the EHS liabilities of specific acquisitions. For example, evidence of past contamination is sought in property acquisitions, and the health and safety implications of specific product lines may be assessed in business acquisitions. EHS managers are generally not involved in assessing the impact of environmental practices that are thought to be long-term or have far-reaching implications. For example, no EHS manager interviewed had been asked to postulate global warming implications during a routine acquisition. Yet these broader implications should be considered by the board, stockholders, and top management to determine what lines of business the company wishes to pursue. Of particular interest due to the current political situation in Eastern Europe and other areas with developing economies, this dilemma is most acute where lack of jobs and available natural resources force unwise short-term utilization of specific high-polluting energy sources.

Reorient Company Production Processes to Waste Minimization and Product Responsibility Goals

At best, corporate EHS managers can be technical and management advisors to senior management and the board in this goal area. In some cases, EHS people have been catalysts for major change in basic processes and product choices, but these have been the exception rather than the rule. Companies that choose this goal now may be ahead in the long run as stockholders, consumers, and communities add their weight to the equation.

Direct Corporate Resources to Infrastructure Development, Training, and Technology Transfer in Developing Countries, Once Basic Standards Have Been Achieved

Large MNCs have resources and options for allocating them to "going the extra mile" to meet stringent standards of OECD countries in all of the host countries in which they operate. A portion of these resources might be better spent in working with the governments and communities in which facilities are located.

The foregoing discussion has attempted to set forth the range of elements that (1) constitute MNCs' environment, health, and safety programs, and (2) drive these programs. The case study discussion presented in the following chapter provides detail on how the elements fit together and are made operational in different companies.

NOTES

1. EHS program components are defined as activities, some of which are highly tangible (a worker training program or an annual auditing review) and some of which are less definitive (management commitment), that taken together become the means by which a company manages environmental issues.

2. Frank B. Friedman, *Practical Guide to Environmental Management*, (Washington, D.C.: Environmental Law Institute, 1988), pp. 27–28.

3. Arthur D. Little, *Environmental, Health, and Safety Policies: Current Practices and Future Trends* (Cambridge, Mass.: Arthur D. Little, 1988), p. 21.

4. Roger Kasperson et al., *Corporate Management of Health and Safely Hazards: A Comparison of Current Practice* (Boulder: Westview Press, 1988), p. 124.

5. "Allied-Signal's Plant Surveillance Program," *Environmental Manager* (1) 9 (1990): 1.

6. Correspondence with EHS director at Consumer Products, June 1990.

7. "A Tool to Increase Competitive Advantage: Environmental, Health and Safety Regulations," *EHS News* (1) 1 (Fall 1989) (Fairfield, Conn.: General Electric Corporation).

8. United Nations Environment Programme, *APELL, Awareness and Preparedness for Emergencies at Local Level*, UN Pub. Sales No. E.88.III.D.3, ISBN 92 807 1183 0 (Paris, France: United Nations Environment Programme Industry and Environment Office, 1988).

9. These were shown to be statistically significant differences as follows:

monitoring compliance: $\chi^2 = 4.7$, df $= 1$, p $\leq$ .03
reviewing capital expenditures: $\chi^2 = 5.2$, df $= 1$, p $\leq$ .02
negotiating with government: $\chi^2 = 5.0$, df $= 1$, p $\leq$ .02

6

Environment, Health, and Safety Programs in Case Companies: *Oil and Gas, Household Products,* and *Pulp and Paper*

The case studies of the following three companies emphasize the operational and program component aspect of their headquarters and domestic production facilities. These cases are designed to complement the material presented in Chapters 1 through 5. This material is not intended to be a comprehensive analysis of a specific company or judgement as to which corporations' programs are the best.

CASE STUDY OF *OIL AND GAS*

Overview

Oil and Gas explores for, extracts, develops, produces, and markets crude oil, natural gas, and coal; and it manufactures and distributes industrial and specialty chemicals.

This case study examines the environment, health, and safety principles and practices of Oil and Gas. The company is organized as a holding company, with its various industry groups conducting the operations. This case write-up focuses on the Chemical Industry Group (CIG). Our interviews took place at Oil and Gas corporate headquarters, a CIG manufacturing facility in the United States, and CIG subsidiaries in Brazil. Individuals from the Chemical Industry Group Technical Center provided frequent and valuable input.

Organization

Oil and Gas conducts its operations principally through five subsidiaries representing distinct commodities. These are referred to as industry groups. Headquarters management describes its organizational approach as "establishing a framework at corporate headquarters with the implementation occurring at the industry group level. We innovate and give the concepts and approaches to the industry groups.''

Reorganization in the late 1980s included financial restructuring with the objective of disposing of businesses and assets not deemed essential to the corporation's core business in order to increase earnings. Over the past five to six years the company has pursued a strategy of shifting its focus from largely foreign operations to business based in the United States. In 1987 Oil and Gas combined its international and domestic oil and gas operations.

Corporate Culture

The company has an image and culture that is relatively youthful and flamboyant. With a history of free-wheeling oil and gas exploration and aggressive company acquisitions, the separate industry groups have an unusual amount of autonomy. The individuals interviewed at corporate headquarters expressed a belief that each of Oil and Gas's principal subsidiaries are among the world's industry leaders in their field in terms of resource size and quality of management.

Corporate culture, with particular regard to environmental issues, has been strongly affected by the company's prior ownership of a major Superfund site. Corporate culture is further described by company personnel as "being motivated by a moral responsibility and, as a close second, motivated by avoiding liabilities for the company." One of the people interviewed at Oil and Gas observed "that the cost to do things right is not that high."

Environment, Health, and Safety: Policies and Programs

The environmental program[1] at Oil and Gas is described by corporate officials as a multidisciplinary and multidimensional program. Its various components take the form of some very tangible initiatives, such as a computerized networking capacity, and some less tangible efforts, such as informal review procedures. However, a large part of what the company refers to as a "program" is also an attitude. The attitude is one in which efforts are directed toward institutionalizing EHS issues. Corporate staff said that environmental protection is "just good engineering" and that these issues should "just be routine." Furthermore, "occupational health and environmental protection should not be separate issues, they should be well integrated into all decision making."

The company experienced a major environmental trauma in the 1970s arising from the acquisition of a company heavily involved in the use and disposal of toxic chemicals. According to corporate executives, these activities acted as a catalyst to formalize and accelerate an environmental program, the core of which included the following recommendations: (1) formalize and tighten an environmental policy; (2) develop and implement a computerized information tracking system; (3) develop an assessment program; and (4) hire a lawyer.

The corporate HSE program is financed independently from industry group operations. There are no chargebacks. Corporate HSE staff feel that if the industry groups or facilities are charged for their services, there is a disincentive for them to be involved. As a senior environmental staff person said, "Why play with funny money?"

At corporate headquarters there is a small environmental staff consisting of a senior consultant (formerly the chief environmental officer) with extensive experience in both line and staff functions in the company, and a staff person with experience in both another multinational company and in environmental consulting. The senior health and safety executives report to a vice president of health, safety, and environment. The vice president reports to the general counsel, who is a member of the board of directors. This core staff sees itself as being organized to step back and look toward long-range planning and implementation; to undertake assessments; to encourage long-range problem solving; and to ensure that the industry groups have the necessary personnel and tools to do their jobs.

Corporate headquarters HSE staff desire to maintain an informal relationship with the industry groups. They feel that the top environmental manager at the industry group should be not more than two slots down from top management. They try to provide some latitude in reporting but attempt not to have HSE report through operations, "who usually have other motivations." Corporate staff believe that the real challenge in the headquarters/industry group relationship is in the politics. This was further described by a corporate person: "The key is the way the corporate folks are portrayed. If the system works right, various disciplines and responsibilities ultimately come together at the implementation level. It's on the facility floor that safety, risk, legal, financial, technical, and all the rest all come together."

Those interviewed at headquarters believe that thin staffing combined with decision-making authority allows corporate environmental staff to concentrate on substantive issues and move quickly to confront new problems. "We don't waste time gathering consensus or trying to justify our jobs."

Industry group staff are described as not being treated with an iron fist approach, rather, they are given responsibility and authority to develop their own EHS personnel and tools. This interaction, management explains, is in part a function of the history of the company, which basically grew by acquisition, with each newly owned company maintaining significant autonomy. They also believe that this management approach puts them substantially ahead of most large multinationals who are just now putting similar systems in place.

Corporate Policy

Company policy calls for line managers to be charged with individual responsibility for the environmental performance of their activities. The company's policy also explicitly states: "Every employee is expected to carry out the spirit as well as the letter of this policy." In addition, a cover letter from the chairman states that he expects "full and complete adherence" to the policy. The policy contains a "commitment to conducting all operations, including sale and distribution of products and services, in compliance with the law." Further required is that an environmental program adopt appropriate standards to protect people and the environment in cases in which laws and regulations are inadequate or do not exist. This includes international operations and provides for a standard of protection of health and the environment functionally equivalent to what the company requires at locations in the United States.

Company policy also calls for (1) the use of modern control methods that are technically sound and economically feasible; (2) a basis for an internal compliance system, including an assessment (audit) program; (3) reporting through multiple organizational channels; (4) line responsibility for employee training; (5) encouragement of process innovations and fundamental research; and (6) provision of adequate resources to carry out strong environmental policies.

Reporting and Accountability

Each industry group has a staff environment department that is charged with defining areas of responsibility; identifying programs and issues; and recommending corrective actions. Although the corporate staff views itself in a "coaching" or "holding company" relationship with the industry groups, there are written procedures that define relationships among the corporate headquarters HSE staff and their counterparts in the field. These procedures are designed to give corporate staff review and comment responsibilities for industry groups' actions, including concurrence prior to appointing any industry group HSE chiefs.

The line managers in the industry groups are accountable to the industry group chiefs for HSE matters, with a "dotted line" relationship to industry group HSE staff, who have a "dotted line" relationship to corporate HSE staff.

Although they believe that the most critical concept in environmental management is that "what you don't know *will* hurt you," corporate staff have put a "management by exception" information system in place. This system attempts to recognize industry group sensitivities concerning corporate headquarters overreach by requiring only information that is significant and that concerns exceptions to day-to-day routines. The system is designed to provide common data for both the corporate HSE and the legal departments.

The system is set up company-wide on the mainframe computer system (although communication from overseas is not always by computer). The intention is to require reporting of only "significant matters," "excursions," and "reportable incidents." The system allows development of a record on any specific issue, which can be retrieved "only" by the responsible facility, the industry group, and the corporate environmental and legal departments. The reporting system was described as flowing from the facility level simultaneously to the industry group and corporate headquarters via the computerized management system.

The intent of this program is to establish patterns and obtain accountability. Since its implementation in the early 1980s, the number of reportable incidents has been reduced by two orders of magnitude, according to corporate HSE staff. Although the computerized program attempts to quantify the number of incidents and excursions, the focus is not on the amount of fines in dollars but on the number of occurrences.

Procedures call for response to proposed and newly enacted laws and regulations and reporting "on a timely basis." Guidelines govern employee awareness and training and the implementation of environmental assessment programs. Guidelines also cover information to be included in requests for significant expenditures and reporting of "substantial risk" under the Toxic Substances Control Act.

Audits

Oil and Gas prefers to call this function "assessment" to avoid confusion with financial audits, which they feel are conducted under circumstances that allow comparisons and judgements. In Oil and Gas's view, it is important to have the audit run by someone from the industry group level. Furthermore, corporate staff say an assessment is more useful as an operational and management tool. Although there are advocates among environmental professionals for a more structured approach, Oil and Gas prefers to keep the process more fluid. They focus their assessments on "systems and major flaws." Corporate headquarters HSE staff believe that this allows for more frequent and more cost-effective reviews. Action plans with target dates are developed to rectify any environmental issues that are discovered.

Corporate HSE staff prepared a guidance document and periodically update it with additional memoranda. Industry group assessments "must meet at least" the minimum criteria specified. Environmental assessments are reviewed by staff not associated with manufacturing operations and are critiqued by the corporate group. The health, safety, and environment committee of the board of directors, the senior general counsel, and the executive vice president for operations review both the assessments and the corporate HSE staff critiques. The program "should make no distinctions between domestic and international operations." An attorney reviews the assessment team's preliminary reports before they are issued in the final form to be sure that items that are or should be subject to attorney-client or work-product privileges are adequately protected.

Acquisitions and Capital Funds Request

Corporate headquarters HSE staff see themselves as having made valuable contributions to the company through their ability to evaluate acquisitions. The sign-off procedure involves three entities: the HSE staff and counsel review, financial review, and an operations analysis review.

Corporate policy with regard to acquisitions is addressed in a written policy: "Compliance with environmental laws and regulations is a matter of highest priority for Oil and Gas's management, not only with respect to existing operations but also as an integral part of its planning for future growth."

The company's board of directors requires that all requests for capital funds be accompanied by an HSE review. These reviews are conducted by corporate technical and legal staff in consultation with industry group personnel. The intent is to avoid problems in plant expansion or modification activities, and to endorse the positive benefits of needed HSE capital expenditures. As was further explained by HSE staff, "The reviews do not entail 'go or no go' recommendations based solely on HSE considerations, but attempt to spell out quite carefully the costs of avoiding future HSE liabilities in the context of the overall business deal."

Training and Certification

Oil and Gas's training programs are described as "modeled and combined with safety awareness programs that have had a long history of implementation and acceptance." Special training programs exist for members of individual facility assessment teams. The staff we interviewed portrayed themselves as a model in the health, safety, and environment arena, yet they expressed a belief that they still have a lot to do toward program improvement. As a senior executive said, "There is no such thing as an accident."

Planning

The current HSE vice president initiated a planning procedure in the early 1980s. It consists of a document that is a compilation of items considered important based on assessment findings; entries in the computerized environmental management and reporting system; and emerging legislative, regulatory publicity, and expert staff issues. The format includes objectives, approaches to be taken, responsibilities, and target dates for both corporate and industry group personnel. The document is updated and circulated for review and comment by the industry groups approximately twice a year.

Since the information is included in reports to senior management and the board of directors, corporate headquarters HSE staff feels that this process is a major lever to initiate and force action. More recently this had been summarized into 12 goals with one- and five-year objectives that are aimed at "the development of preventative programs and procedures designed to minimize liability." Those interviewed at corporate HSE noted that the details of program implementation to achieve the goals are the responsibility of the industry groups, with corporate headquarters being more interested in a strong emphasis on the concepts and initiatives taken by the industry groups.

Equivalency and Technology Transfer

Environmental program guidelines at Oil and Gas call for implementation, at all international operations, of a standard of protection for human health and the environment that is "functionally equivalent" to what is required at locations in the United States, or to provide documentation to the contrary. As described by corporate HSE staff:

> [O]n paper, the first step in implementing a functional equivalency policy is to under-take research on local requirements. This company has found that when these re-quirements are placed in tabular form next to the requirements for a representative facility in the U.S., the differences are usually minimal. It is hard to argue that health, safety, and environmental requirements should not at least meet the re-quirements of the host country.

The corporate HSE staff believes that there are a number of ways overseas to accomplish the goals of U.S. environmental legislation, particularly the re-quirements that are technology-based. A company practice was described as bring-ing in a "responsible expert" to document in the permanent records of the cor-poration the basis for concluding that company actions afford equivalent protec-tion compatible with the intent of the policy. Staff claim that this approach has caused a minimum number of requests for exceptions to the policy. An HSE ex-ecutive explained:

> It is rare when environmental issues and regulatory issues become a leveraging fac-tor when siting and operating a facility. The key factors are markets and access to raw materials. Most of our headaches come from the locations where we have fairly new operations. In the locations where we have been operating for a while, we have established positive working relationships.

Environmental personnel at headquarters discussed the tremendous impact that the Bhopal disaster had on certain policy initiatives. As a result of this industrial crisis, Oil and Gas developed an edict "ensuring worldwide protection from all impacts from an Oil and Gas facility."

Oil and Gas corporate staff indicated that health and safety programs can be improved at overseas facilities; however, one stumbling block is the difficulty of obtaining appropriate equipment. Oil and Gas indicated that this is due to a variety of different circumstances—from snags at Customs to difficulty obtain-ing locally the proper sampling and analytical equipment. Attempts are made to hire local consultants to perform these services. Industrial hygiene testing is the responsibility of the individual industry groups and is handled locally. Training of local personnel is said to be provided locally, thus insuring that "proper techni-ques, adapted to local conditions, are used."

Corporate staff discussed the success of technology transfer occurring through the movement of personnel both within the company and among other multinationals.

With regard to facility operations overseas: "Our management design is that the expatriate manager will develop a subordinate deputy's technical expertise so that the [local] deputy can take over the job responsibilities within a five-year period."

Facility Home Country: United States

History of Facility and Location

The Oil and Gas facility we visited is located in an area dominated by chemical manufacturing plants. There are 10 major chemical production and distribution operations within immediate proximity to the site under study. This plant is a relatively recent acquisition, coming under the current ownership and management less than four years ago. The 25-year-old facility operates 24 hours a day, 365 days a year. There are 170 employees, 47 of whom are management and administrative personnel.

Oil and Gas views itself as a respected employer in the area. Recently the company placed an advertisement for new hourly help, and it generated 70 applications. It was emphasized that in a state with very low unemployment, these were not applications from people who lack alternatives. "We are viewed as one of the better places to work."

Products and Process

This plant produces approximately 380 tons per day of chlorine, placing it in the small to medium size category of chlorine production. Chlorine is a highly reactive chemical element; if it is not handled with extreme caution, it can cause nose, windpipe, and lung irritation. Heavy concentrations can cause death.[2]

The facility uses the mercury cell process to produce chlorine and the co-products of caustic soda and hydrogen. Chlorine is produced by the electrolysis of brine (salt dissolved in water). Other processes that are used to produce chlorine are the diaphragm cell and the membrane cell. The latter two processes require less energy per unit of product, but the mercury cell is the only process that produces highly pure caustic soda that is marketable without further processing. The diaphragm cell is predominant in North America—about three-fourths of the continent's capacity.[3]

Environment, Health, and Safety Issues at the Plant

The environmental manager described his responsibilities as (1) maintaining compliance; (2) fulfilling permitting requirements; and (3) monitoring and anticipating regulation changes. "Much of my time is spent looking down the road, identifying issues and priorities."

A primary issue for plant operators is hazardous waste generation, reduction, transport, and disposal. Facility management projected an ability to significantly reduce hazardous waste over the previous years' generation. Facility management discussed projects currently under way involving on-site treatment to render waste nonhazardous. Facility management believed headquarters would support waste reduction initiatives by ensuring availability of capital for improvements.

Organizational Structure

This facility is organized such that the plant manager has several direct reports: operations, controller, safety, human resources, technical, maintenance, and purchasing managers. The technical manager supervises the process engineers, the environmental engineer, and the laboratory supervisor. The operations manager has direct reports covering shipping, customer relations, shift supervisors, and production and process supervisors. The organizational structure in place attempts to provide a series of cross-checks on issues of overlapping concern.

Management Philosophy and Style

The management of this facility expressed the influence, technically and philosophically, of corporate headquarters. Establishing environmental goals, for example, is a process mutually agreed upon by corporate headquarters, industry group headquarters, and the facility. The specific technical approaches are implemented at the operations level with technical input provided by the industry group technicians.

Environment, Health, and Safety Programs

Training in the area of safety and risk issues is a primary responsibility of the facility operations manager. The operations staff indicated that each facility within the chemical division handles procedures differently; therefore, process and personnel alterations were carefully monitored, with the appropriate training taking place concurrently.

The safety staff at the plant are responsible for running drills. The procedure calls for drills to be run once a month to test a master plan, with an occasional emphasis on a specific emergency aspect, such as the handling of injured people or managing media interviews. These drills, along with "spot checks," can and have been initiated by industry group headquarters.

Facility personnel discussed a corporation-wide waste minimization program currently under way. This facility staff expressed confidence that they have the management and mechanisms in place to achieve their reduction goal.

HSE staff discussed an annual environmental audit. The audit is described as a performance review as well as an opportunity to track issues and determine dates for completion. A separate process also sets goals for controllable incidents, for example, permit excursion and waste water discharge.

Technical as well as management personnel participate in training programs conducted through the industry group headquarters. These vary in program focus as well as frequency and duration. There are two manufacturing meetings per year that last a week each. Cost, safety, environmental regulations, long-range planning, and other items are included on an agenda reflecting major responsibilities of a plant manager. Some specific operations issues can be addressed at these company-wide meetings. For example, mercury contamination of water presents significant problems, and these meetings provide an opportunity for all of Oil and Gas's plants facing the problem to collaborate and exchange information on control methods.

The plant participates in a hazards review program. This is initiated at the industry group level and is implemented on a facility-by-facility basis. The program examines in great detail specific HSE issues. These issues are broken down and specific responsibilities are assigned to a program manager. This individual is expected to collect all relevant procedural materials, probe for deviations, and develop a standard method to achieve the best and safest manner to address the issue.

Reporting Mechanisms

The plant manager discussed writing a monthly report to the industry group headquarters. The report includes environment, health, and safety concerns including excursions from permitted parameters, cost issues, and other items that are delineated in a corporate operations manual.

Community Relations

Oil and Gas sees itself as very proactive in the area of public and community relations. Because the facility location is highly industrial, the majority of its initiatives have involved issues of particular interest to other industries. This facility organized a press conference around SARA Title III in an attempt to establish worker and media communication.

Other examples of Oil and Gas's role in the community include hosting local industry organization meetings and involvement in local and state industry trade associations. "This plant has always provided the leadership to do things." It was clear from those interviewed that the tradition of local leadership went back to the previous ownership and management. The current activities are very similar to those carried out prior to acquisition.

Facility personnel rely on national trade associations, such as the Chemical Manufacturers Association (CMA), as an information source. CMA is putting together state advocacy organizations to ensure representation of the chemical industry. Plant personnel intend to play a leadership role in the local effort related to CMA activities, in particular the Chemical Accident and Emergency Response Program, which was initiated by this facility in 1985.

CASE STUDY OF *HOUSEHOLD PRODUCTS*

Overview

The company we are calling Household Products is a manufacturer and distributor of household goods including personal care products such as deodorant and toothpaste, and laundry and cleaning products. It employs over 75,000 people with operations in 50 countries totaling more than $19 billion in sales.

In 1987 the company began a restructuring program aimed at making product introduction more agile and avoiding intra-company competition. Category business units were established as core profit centers throughout the company's domestic operations. All brands in a particular business category, such as toothpaste, are grouped organizationally in the same unit. In addition, manufacturing, engineering, purchasing, and distribution personnel have been grouped together into new product supply units to better meet the needs of each business category.

The 1987 restructuring is seen by analysts as part of a continuing drive to streamline management of a diverse company. Household Products calls this the biggest management change in more than 30 years. The new organization is intended to minimize the conflicts and inefficiencies that occurred as brand managers competed for corporate resources, and to emphasize how brands can work together.

According to the individuals we interviewed, the company is steeped in a strong tradition of values and high moral standards. Contemporary corporate attitude is strongly oriented in a business, sales, and marketing direction. A predominant attitude expressed throughout the interviews was that values and principles drive the company's environment, health, and safety programs, not compliance or regulatory issues.

Environment, Process Safety, and Health and Safety: Policies and Programs

The Environmental Control Division (ECD) serves a staff function for the management of environmental issues at plants worldwide. The mandate of this division is described as a monitoring, assessing, and measuring function along with technical consulting and problem solving. Furthermore, ECD sees itself as carrying out the corporate environmental consciousness. Its primary area of responsibility is manufacturing process and impacts as opposed to product safety issues. The majority of the 30 individuals at the ECD have formal training in engineering.

Household Products has a written corporate policy that addresses environment, health, and safety issues. According to headquarters staff, this policy was rewritten in 1990 in an attempt to integrate all aspects of the corporation's business, from product development, engineering, packaging, manufacturing, marketing, and advertising to the consumers' end use and disposal. A multifunctional team representing all disciplines has been formed to coordinate the implementation of this policy. The ECD represents engineering and manufacturing on that team.

According to a corporate environmental official, "The policy is meant to be truly international in its implications and has been translated into 27 languages, as has a video presentation on what the company is doing on the environment." A further component of this policy has been the appointment of a manager for environmental corporate coordination reporting directly to the chief executive officer (CEO). The position is responsible for coordination of environmental issues across all aspects of the businesses, as well as supporting functions such as corporate environmental, legal, and advertising staff.

The management of environmental programs and resources for manufacturing from the corporate perspective is highly decentralized. The ECD has developed a system that is designed to funnel expertise to the plant locations. Management encourages reciprocity, such that plant personnel are technically able to interchange with headquarters personnel. As an ECD staff person said, "We are trying to minimize the perception that everyone from headquarters is the CEO." Furthermore, ECD is "trying hard to stay away from a 'we/they' or 'us/them' system in regard to the relationship with the facilities. We want to see our people all over the world trained as part of a network."

Both corporate staff and facility engineers are expected to have skills and training enabling them to deal with a range of environment-related issues, including technical; managerial; public relations; legal; and policy issues. Corporate staff describe a continuing effort to encourage and support autonomy and technical expertise at the plants.

The costs of the ECD are covered by the business operations. There is a one-third direct charge from ECD to the business divisions for their services, and two-thirds of the ECD budget is allocated to them on a sales volume basis.

The ECD sees itself as playing a large and significant role in the U.S. environmental regulatory arena. They expressed the view that "it's better to get ahead of the regulatory curve by participating as a driving force. We shouldn't wait for laws, we all have a responsibility to the environment—it's just a matter of making some hard business decisions."

The process safety program is focused on providing the technical safety apparatus, training, hardware, maintenance, and inspection programs for Household Products facilities. The corporate headquarters office of process safety is responsible for preventing accidents causing any interruption of production, loss of property, or injury at all manufacturing sites. The three major causes of accidents involve fire, explosion, or excess pressures. The process safety issues for the European plants are the responsibility of a Household Products European office.

Process safety at Household Products was described by the program director as a "high-commitment work system," the basis of which involves written "Plant Process Safety Rating Guidelines." The six functional areas are rated either superior, adequate, or marginal. The areas are (1) control of process changes;

(2) safety of plant design; (3) maintenance and operation of process safety equipment; (4) plant safety—as built; (5) training and behaviors; and (6) continuous process safety improvement.

Paralleling in many respects the philosophy of the Environmental Control Division, the Process Safety Division has a strong commitment to the headquarters management being well versed in plant-level issues with frequent hands-on experience. Process safety personnel discussed a management approach that looks at the plant as the customer or consumer of headquarters' expertise.

Process safety, as a subdiscipline of environmental or engineering applications, is not as heavily regulated. It was further explained that "the hardware is essentially the same as it was 40 years ago; it's the science and application that has advanced rapidly. I can't recall very many hardware issues; for the most part we focus on the software, people issues." The result is that process safety programs are designed to rely on "a lot of selling and promoting of risk management values."

Industrial Health and Safety (IHS) at the corporate level is responsible for the policy, standard setting, and technology transfer of occupational health issues. Workmen's compensation, health care cost containment, and employee assistance programs are other areas of responsibility. In developing programs to address worker issues, interviews stressed a strong emphasis on values and principles as opposed to programs being compliance driven. An IHS director claimed; "average is never OK around here," and "if the principles are there, execution will follow." The IHS programs are oriented toward a management systems approach to implementation. "Management systems have to be in place in order to have the health and safety activities happen," said an IHS employee.

In discussing the health of employees, our interviews with IHS personnel raised the issue of how a negative health effect of a product on a Household Products worker would impact the company. As corporate staff elaborated, "In a manufacturing process, if one of our brand names has any medical implications, we'll spend whatever's necessary to clean it up." Further discussion took place involving the significant resources used by Household Products to research health impacts through worldwide epidemiological studies on workers.

Health and safety standards at Household Products plants located abroad tend to be comparable to U.S. standards, an IHS director explained. A plant manager at a foreign facility can modify an operating standard only to make it more stringent.

Reporting and Accountability

The Environmental Control Division, Process Safety Division, and Industrial Health and Safety all have several formalized reporting programs designed to be implemented worldwide. They vary in the level of detail provided and the reporting route. However, headquarters sees the common element of performance improvement in each program. Current programs discussed with the ECD include:

- Environmental incident reporting program—identifies what needs to be reported, and reporting procedures, and provides the data for semi-annual summaries on incident performance;

- Environmental compliance issue program—division-specific management of compliance issues designed to expedite their resolution; statistical methods are used to document capable systems;

- SARA Title III compliance program—division-wide program to communicate federal requirements to U.S.-based facilities; and

- Underground Storage Tank (UST) risk reduction program—requires each site to have a management plan for its USTs to minimize future liability and ensure compliance; periodic testing is included.

Reporting and accountability programs were discussed, as the following sections indicate.

Audits

Environmental audits and ratings are two methods used to measure performance. In January 1988, ECD initiated a revision that it believes improves the auditing process by including an objective rating system. The program requires annual audits and ratings of the environmental systems at each site. The audit focus is on continuous improvement. ECD conducts audits/ratings once every one to four years and requires self-audits by the sites in the other years. Although headquarters requires that an audit take place annually, ECD personnel are not always present for the audit.

The audit team usually consists of the plant manager, a plant manager from another facility, and an individual from the ECD at headquarters who is familiar with the operation. These audits are approached with the philosophy that they are useful as training tools among the various plants. ECD is involved on a rotating basis in order to make cross-business/boundary comparisons. Plants are rated on environmental management system performance in the context of site complexity.

Staff from the Process Safety Office discussed inspections of each facility every two years with a specific emphasis on the six functional areas mentioned earlier. These audits are viewed as products, with a "detailed assessment of key elements."

A system is in place that is designed to allow corporate lawyers an opportunity to review the environmental audits of non-U.S. facilities for content and also the manner in which the information is presented. "We need to see if the liability can be transferred back and hold corporate responsible." Corporate legal offices further expressed the sentiment that they "would prefer that audits weren't done at all."

Acquisitions

Expansion of existing product lines has been the traditional method of Household Products' growth until fairly recently. In the early 1980s, the company began to pursue aggressive growth through large acquisitions. This has presented challenges to the environmental professionals, from a technical as well as a corporate culture perspective. Having long-term loyafty to the company has given common operating ground in dealing with environmental issues. However, new acquisitions present dilemmas such as: "How do you bring new people along who haven't grown up with the Household Products culture?"

The acquisition policy calls for the inclusion of ECD in the review of proposed investments. This review procedure focuses on "deal stoppers" typically in the million dollar–plus range. As experience with the acquisition process grows, its use is expanding to international locations. Recent acquisition assessments have been done at multiple sites in Europe, Canada, and South America. Yet there are situations in which ECD review is not included, or timing does not allow for the review to be as thorough as the ECD professionals would like it to be. As an ECD employee indicated, "the words are there, but still, there's so much we're buying, we just can't get them all." Corporate legal offices expressed similar experiences in that a system was in place to examine legal implications, but it did not always operate as efficiently or as thoroughly as possible.

Environmental Business and Risk Management Teams

Each business division and manufacturing operation within that division has varying needs in terms of environmental expertise. Environmental business and risk management teams have been set up to determine needs and assist with technical matters. These teams meet four times a year to coordinate their activities. Membership consists of individuals from the business division, the specific manufacturing operation, and ECD, with additional input from corporate business, finance, or legal offices.

Training and Certification

A formal training/certification program attempts to ensure that site environmental and process safety managers have the minimum technical skills to accomplish their job. This program was initiated in October 1986 by the ECD and has been tailored to meet specific division needs since that time. Each site is said to require a trained and certified plant environmental manager or contact and a process safety technician. A technical safety workshop is scheduled every two years. This is described as a two- to three-day program with a significant computerized networking component. The international divisions of Household Products send their people to be trained at headquarters and training facilities. The training program is characterized as a "behavior-based systems approach."

A manual called "International Plant Environmental Contact Training Book" has been developed. This reference includes, in significant detail, programs, policies, and procedures for the international plant management in addressing environment, health, and safety issues. The manual is written in English with accompanying training tapes translated into the language of the country of operation. These tapes are developed to facilitate training aimed at hourly workers.

Facility Home Country: United States

Overview

The facility visit consisted of interviews with plant personnel in the spring of 1989. The location of this study company's facility is in close proximity to company headquarters—the same metropolitan region. The manufacturing site represented three separate production facilities, one of which was the oldest the company owns, built in the late 1800s. There are a total of 1,300 employees on the site.

The manufacturing operations include production of detergents, bar soaps, and food products. The process toured for this project, because it is replicated at an overseas location, was the detergent production operation. This process involves receiving and storing raw materials; paste making; mixing and pumping; spray drying; cooling, screening, and mixing; and packaging of synthetic detergents.

Environment, Health, and Safety Issues at the Plant

According to the environment control manager, his responsibilities involve the following: permit renewals; waste water compliance issues; odor complaints; spill response plans; underground storage tank removal program; process system improvements; paperwork and report writing, both internally and to state and federal environmental regulators; and community relations issues.

The process safety focus at the plant is on the training of process engineers and on the design and maintenance of equipment. Relief valves, pressure vacuums, secure tank covers, and lids were issues of a routine nature.

Organizational Structure

The facility is organized so that the environmental control manager reports to the plant engineer, who in turn reports to the plant manager. The plant manager is responsible to a division manager who typically is in charge of two to four different facilities. Each plant has a designated contact within ECD to facilitate communication and resource decisions.

Management Philosophy

Management philosophy and style are described, by those interviewed, as "now management." This was defined as an approach less wrought with cumbersome bureaucratic apparatus, focusing on swift decision making at the lowest level possible. One individual at the plant further explained the management style as "less dictatorial in implementation, with more emphasis on performance rather than prescription.'' Innovative and risk-taking actions were said to be encouraged at the manufacturing plant level.

Environment, Health, and Safety Programs

The major environmental issues at the plant level are addressed in the following six programs:

1. air emissions—major components are inventories and periodic compliance testing of emissions sources;
2. hazardous waste—major component is to utilize a manifest system to manage the waste generated and assure proper disposal;
3. polychlorinated biphenyl (PCB) elimination—locate and identify PCBs and address through removal from site to incinerator;
4. groundwater protection—address groundwater contamination from two sources, removal of underground storage tanks and prevention of contamination through improvements in dike management;
5. waste water and sewage—maintain compliance with local treatment facility requirements;
6. dust control—minimize dust level to protect workers and comply with OSHA regulations.

Although these programs are based around general compliance and regulatory areas, the environmental control manager felt strongly that ''we're driven by what's right, and compliance is secondary.''

The process safety manager at the facility level stressed the importance of training and particularly the re-training of personnel when there is a change in a process. The six functional areas of the Plant Process Safety Rating Guidelines provide the basis for the process safety programs.

Reporting Mechanisms

The facility environmental control manager has a dual reporting relationship: to both the plant engineer and a ''dotted line'' to corporate environmental control.

The facility environmental control manager contributes to a routine public affairs report. Any community issues that involve EHS are funneled to the site public affairs manager. This individual sends information to the corporate public affairs group. The report is then used to position upper management, particularly when a vulnerable or action-oriented issue has surfaced, and it is described by those interviewed as a ''mostly proactive document.''

The facility environmental control manager is responsible for routine memos as well as monthly departmental reports. These are not mandatory and are viewed internally as " 'a keep me informed program,' not necessarily regimented." The plant communications are supplemented by the annual reporting of progress on each of the environmental management systems tracked at the corporate level by ECD.

Community Relations

In the corporate hometown, the individuals we interviewed referred to the reputation of the company as "an excellent corporate citizen." They have a generous charitable giving program, with recent endowments to local and state educational institutions. Volunteerism and support for minority economic development programs are strongly encouraged.[4]

CASE STUDY OF *PULP AND PAPER*

Overview

Pulp and Paper is principally engaged in growing and harvesting timber and manufacturing, distributing, and selling a wide variety of forest and paper products. The company has expanded and diversified over the years to include real estate ventures and the marketing of investment products. The company has operations around the world, with most of its significant manufacturing and production operations in North America. Overseas activities consist primarily of sales, service, transportation, and container operations. Pulp and Paper owns or has harvesting lease rights on over 25 million acres of land worldwide.

The following discussion is based on company documents, library research, and interviews with corporate officials and facility management and staff. The facility tours and interviews took place at a pulp and paper mill located in a state neighboring the corporate headquarters, and the second facility tour, the "foreign facility," was a pulp mill located in western Canada.

Organization

Pulp and Paper is organized around a central holding company and three principal business groupings: P&P Forest, P&P Paper, and P&P Diversified. During the particularly competitive business climate of the early and mid-1980s, Pulp and Paper went through a period of consolidation and structural reorganization. Many less profitable product lines were sold off and overseas installations and property were divested. The advent of a strong dollar and a flooded forest products market further exacerbated poor business conditions. Personnel cutbacks shrunk the company staff by approximately 10,000 jobs. Currently, worldwide Pulp and Paper has increased its employment from the leaner early 1980s, and it employs approximately 45,000 people in businesses with assets over $7 billion.

The central holding company operates out of corporate headquarters in the Pacific Northwest. Headquarters provides numerous services to the various business groups. These services are available on a chargeback basis to the businesses. The chargeback system was instituted at the time the company underwent structural reorganization in the early 1980s. It began as a resource efficiency program to provide services in research and development, legal affairs, government relations, environmental affairs, and health and safety. The intent was to centralize these areas of expertise and minimize duplication of services.

P&P Forest manages Pulp and Paper's 6 million acres of commercial forest land in the United States. This is the largest volume of privately held timber land in the world. The company produces both hardwood and softwood lumber. Its major products include timber, logs, chips, plywood, particleboard, hardboard, other composite panels, and laminated beams and decking.

P&P Paper has approximately 5 million tons of annual primary capacity. Its production facilities serve market demands for papergrade kraft, fluff for towels, disposable diapers, and sanitary napkins, and specially sulfite grades for lacquers, artificial silks, and photographic papers. Other products include container board for industrial and agricultural packaging, printing and writing papers, newsprint, bleached paperboard, and containers for liquids.

P&P Diversified includes a real estate company, a holding company for regional real estate building and development firms and for an urban commercial real estate joint venture company. P&P Diversified produces personal care products and disposable diapers, and it markets nursery and garden supplies.

Pulp and Paper, although operationally an international company, perceives itself more as an exclusively North American firm. The majority of its most significant interests are in the United States and Canada. Since the company has limited operations internationally in terms of manufacturing and production, the international management focus is primarily along the sales, service, transportation, and container lines of business. The exception to this is in Canada, where there are operations and holdings including pulp and saw mills.

Corporate Culture

In comparing itself to others in the business, the company views itself as being in the forefront of environmental management. A sense of commitment and loyalty to the company and the environment seem to exist at Pulp and Paper. This was apparent during interviews as well as appearing frequently in company literature. Some of this can be attributed to a long-standing, highly visible sense of family tradition and ownership. We were told: "Pulp and Paper has maintained a commitment to long-term planning and business strategies, maintaining timber as a crop rather than a resource to be exploited and then abandoned."

There is strong evidence of the personal commitment to the environment by the company's president and chief executive officer. This individual has taken the position that the burden of proof that an action may or may not be harmful should be on the company, not on the environmental activist. He also has stated on numerous occasions that "the strength of each individual employee's own ethical system will determine the good or bad environmental practices of a company."

Environment, Health, and Safety: Policies and Programs

Pulp and Paper's environmental issues are managed through the environmental, energy, and government affairs group (EEGAG). This group is divided among five regional environmental affairs managers (REAM West, REAM East, REAM South, REAM Northeast, and REAM Midwest). Each REAM team is managed as a one-person operation, coordinating with resources at corporate headquarters as well as with environmental personnel at the facilities in the respective regions. The professional training of the REAM teams is strongly oriented toward the scientific and technical disciplines. Each REAM team reports to the vice president of the environmental, energy, and government affairs group.

The magnitude of impacts on the environment inherent to the forest products/pulp and paper industry have resulted in very regulatory conscious personnel. The environmental staff at headquarters describe themselves as being in a coaching and advisory position. This function is carried out not only through interactions with the facilities but also by working with and providing technical expertise to the regulators. The staff cutbacks in the early 1980s forced serious evaluation of personnel resources. The headquarters environmental staff now see themselves as running a "lean and highly efficient team." They operate on a company philosophy that encourages driving decisions to the lowest possible level in the hierarchy. The company is moving toward incorporation of environmental, health, and safety issues into the overall decision-making network.

The corporate headquarters safety and health services is organizationally and functionally separate from EEGAG. However, consistent with a decentralized organization, each facility location is staffed with its own environmental and health/safety expertise. The facility professionals will call on the corporate resources in some specialized situations such as an inquiry into chronic or long-term health exposures or health effects.

Pulp and Paper relies on an issues management team to identify, define, and manage corporate environmental concerns. Depending on whether an issue is long term or short term in nature, either an ad hoc or more formalized approach is used. The effectiveness of this program is increased by drawing on expertise from various department managers and encouraging an interdisciplinary approach to problem solving.

The management relationship between corporate headquarters and the various Pulp and Paper operations is one that encourages a high degree of autonomy at the facility level. Each facility operates as a self-sufficient unit, staffed by its own environmental and health/safety experts. The facility professionals utilize the corporate resources through chargeback when their local, in-house capabilities are exceeded. Corporate personnel become directly involved in some issues at the plant level, particularly in dealing with large regulatory and permitting issues.

External professional resources such as trade associations and research councils are used extensively. They are relied upon primarily for political lobbying efforts, technical updates, and issue collaboration with other member companies. External consulting firms are used when corporate services are overextended, or when a local firm is either logistically or financially more competitive.

Corporate Policy

Pulp and Paper has an "Environmental Policy and Statement of Corporate Values" (EPSCV) reflecting one of the most detailed and comprehensive corporate policies available. The EPSCV was first developed in 1971. The process used in developing this policy was to first identify other companies with reputable environmental policies and practices. Then, this material was reviewed and synthesized, and a policy statement was created that reflected the company values. The EPSCV is a comprehensive indicator of the company's attitude and responsibilities. The emphasis is on complying with regulations, integrating environmental concerns into overall decisions, and individual employee responsibility and contributions.

Reporting and Accountability

Audits

After the corporate policy was developed, inquiries by corporate management sought assurances that a mechanism was in place to follow through on the intentions of the EPSCV. Thus, the development of a corporate policy was followed by the structuring and implementation of an audit program.

The environmental audit program has been developed at the corporate level and is in the process of being implemented at the facility level. It is said that this audit program is "driven from the top" and that the CEO wants to be assured a secure level of compliance.

Similar to the process used in developing the environmental policy statement, the audit procedure was developed as a result of extensively surveying the audit procedures used by major chemical companies. The two-part procedure involves first training the audit team, then going through a facility to evaluate and inventory environment-related concerns.

The basic tool in the audit is a manual that compiles major environmental laws in the form of questions and checklists. The manual is given to the facility management personnel, and they review and go through their facility utilizing the information from the manual to key in on issues of concern. Approximately one month later an audit team (consisting of a corporate environmental manager, a manager from a facility of comparable complexity, and the facility environmental personnel) tour the facility and rank concerns on a scale of one, two, or three. Priority one means that a critical condition exists and needs to be addressed immediately; priority two indicates that a concern exits although not of the magnitude or urgency of a priority one; and a priority three means that the issue is more of a maintenance or housekeeping nature.

The audit program is in the implementation stages. Ideally, the plan calls for an audit to be performed at each facility on an annual basis. The corporate managers of this program are optimistic regarding the success of this approach because (1) it involves a team strategy; (2) top management from corporate headquarters as well as the facility play active roles; (3) plans are developed to address outstanding concerns and (4) technical expertise and capital funding sources are provided. The process indicates positive results in moving from a theoretical corporate policy to a tangible, results-oriented program at the facility.

Pulp and Paper headquarters staff envision the audit as a vehicle that initially will serve to counsel and train. The plan is then to move the audit program from a reactive tool to more proactive initiatives.

Facility Home Country: United States

The Pulp and Paper facility we visited in the U.S. had an extensive safety management program directed at its workers. They participate in a "Safety Exchange" on an annual basis. This exchange is similar in concept to the environmental audit, with a stronger emphasis on occupational issues. It provides a "consistent measurement of the priority emphasis" using 14 key elements. These elements include activities such as management plan, emergency procedures, employee orientation and training, and accident investigation.

The facility management demonstrated their commitment to health and safety issues through training and refresher programs for both full-time employees and those hired on a contractual basis. Environment and safety issues are the priority item at each morning briefing session of the day shift. This facility is active in the Pacific Coast Association of Pulp and Paper Manufacturers and holds ranking within the top 10 of a possible 43 in terms of companies' ranking with least injury and illness reports.

Baseline surveys are used for both environmental background and health screening. Health screenings occur at time of employment, and discussions are under way for expanding this procedure to include longer-term health effects studies.

Both the environmental audit and the Safety Exchange use a combination of labor and management as part of the review teams. The audit requires the team members to participate in a training program prior to the field work. This is designed to attempt a well-informed and consistent approach.

An objective of these programs is that education and cooperation be highlighted. Although a certain amount of competition exists among the facilities, there is widespread belief that an improvement at the facility will benefit the entire company.

Pulp and Paper's environmental, health, and safety program is driven by two fundamental sources: a strong corporate desire to adhere to the values in the corporate policy, and the ability to maintain compliance with regulations. There are indications, however, that the company is evolving into more proactive initiatives and views itself as taking a more aggressive leadership role.

NOTES

These cases were prepared to stimulate discussion and are not intended to illustrate either effective or ineffective approaches to the situation presented. The material is based on interviews conducted in a real company; however, fictitious names have been used.

1. The order in which the terms *environment, health,* and *safety* appear differs among companies. Oil and Gas prefers Health, Safety, and Environment (HSE) in identification of its company programs.

2. Chlorine Institute, "Chlorine: A Guide for Journalists," Pamphlet 70 (Washington, D.C.: Chlorine Institute, February 1980).

3. Ibid., p. 70.

4. Council on Economic Priorities, *Rating America's Corporate Conscience* (Reading, Mass.: Addison-Wesley, 1986).

7

International Challenges: Standards for Environment, Health, and Safety

OVERVIEW

How an MNC manages its environment, health, and safety programs in the many different regulatory regimes in which it operates is an issue of emerging importance for corporations. An approach at one extreme is to have each facility simply meet the applicable regulations in the country of operation, and at the opposite extreme to develop uniform corporate standards that meet or exceed the requirements in place in all countries of operation. None of our case companies operate at either extreme; this chapter explores how companies are addressing the challenges associated with establishing comparable operations.

WHOSE STANDARDS?

The setting of EHS standards in multinational operations is challenging from a number of viewpoints. The classic problem is posed as follows: In a given host country, there are 10 facilities owned by eight different MNCs. Each company is owned and headquartered in eight different countries. Each of these countries has different EHS standards. If, on the one hand, each company's home country standard applies in every one of its facilities, host countries are confronted with an impossibly complex regulatory scenario. On the other hand, if (as is often the case) each facility binds itself to comply with the standards of the host country, and standards vary from host country to host country, the multinational company finds itself enforcing different standards all over the world, depending on host country expectations.[1]

In addition, problems of competitive fair play arise when costs to comply with EHS regulations fall harder on one multinational than on another, and when they fall harder on the multinationals than on local firms. One case company summed up its corporate practice as "standards that are practical in the environment in which we operate."

In China, a case company has a joint venture under way with the government. A corporate staff member described conditions as "primitive."

> The deal all along was to build the plant to American standards, not Chinese. It shouldn't have been a problem, but it didn't work. The Chinese made decisions on using their own standards during construction. For example, they did the electrical system using open knife switches. We are now spending about $900,000 to redo the electrical fixtures, and we are negotiating to share the costs 50/50. I just received the technical specifications and I want to be flexible to accommodate local purchase where possible, but there will be cases where I have to insist on U.S. equipment.

Where a corporation has limited control, even basic U.S. procedures can require considerable effort to put in effect. These issues are being discussed and negotiated at the moment throughout the world.

In theory, the promulgation of uniform international EHS standards could be of substantial benefit to MNCs, if the standards were perceived to be reasonable and if enforcement were even. Currently, enforcement is uneven throughout the world, even within and among the industrialized countries. With balanced application of uniform standards, a company could feel secure that competitors were not benefiting by exploiting differences in EHS regulations. This concept is often referred to as "creating a level playing field." However, several issues will need to be addressed and resolved before international standards can be made acceptable to the MNCs.

It has been argued that smaller companies and state-owned companies will never be able to afford the technology and management to satisfy stringent international standards. One possible approach to this challenge is through transfers of technology and management expertise; nongovernment organizations such as the International Environment Bureau and the World Environment Center have begun to establish success records in such transfers from developed country to developing country industries. One corporate expert who participated in a technology transfer mission to India commented: "The World Environment Center program costs [the company] a lot of money, but the company feels it's part of its public service. . . . We see it as part of our effort toward saving the world."

Another challenge is to design a regulatory system that will promote and encourage pollution prevention strategies and nonwaste technologies, with clear advantages for sustainable development, rather than a system that creates a bias toward end-of-pipe solutions.

Even if the foregoing issues could be resolved, there remain significant arguments suggesting that uniform international standards could, under some circumstances, work against sustainable development by forcing inappropriate priorities on developing countries. For example, under uniform international standards, high levels of environmental protection against cancer might be achieved at the expense of basic human needs such as protection from high infant mortality and rampant malnutrition.

The phaseout of CFC production under the Montreal Protocol recognizes the special problems of developing countries by giving them more time to comply, but one can see the limitations of a similar treaty approach if applied to the full range of environmental protection issues as codified in U.S. law.

WHAT INFLUENCES EHS DECISIONS?

In our survey, respondents were asked to indicate the degree of influence a variety of factors had on their companies' EHS decisions in different regions around the world. The factors were as follows: level of government enforcement; liability issues; government laws/regulations; community opinion; customers; and high-visibility accident or release to the environment. The choices for the degree of influence were as follows: very influential, somewhat influential, and not influential. Figures 7.1 and 7.2 illustrate the level of influence of selected factors.

Respondents indicated that the greatest influence on their companies' EHS decisions in the United States is government laws and regulations. In contrast, companies with operations in Africa indicated that concern for high-visibility accidents had greater influence on company decisions there than did government laws and regulations.

Community opinion and customer demands had similar influence on company decisions in the United States, Canada, and Europe; however, customers had greater influence than community opinion in Southeast Asia and Japan, the Middle East, Latin America and Mexico, and Africa.

THE LURE OF INTERNAL STANDARDS

EHS programs at the facilities of most MNCs, including those in this study, are principally "compliance oriented." Such programs are driven by government priorities and are responsive to changes in environmental regulations. T. H. Lafferre has observed that until recently, both environmentalists and industry believed the myth that compliance was the ultimate goal of corporate programs. Describing industry's approach, Lafferre said: "Business strategies began to center around compliance, investments in manufacturing plants centered around compliance, our internal organizations were structured around compliance."[2] However, some companies are developing concepts of environmental stewardship that go beyond a compliance orientation. These efforts take a holistic view of the company's relationship to the environment and incorporate environmental considerations into virtually every aspect of the organization's decision making.[3]

There was a widespread sense among the corporate EHS staff we interviewed that they should be working toward a situation in which there are comparable

Figure 7.1
Influences on EHS Decisions

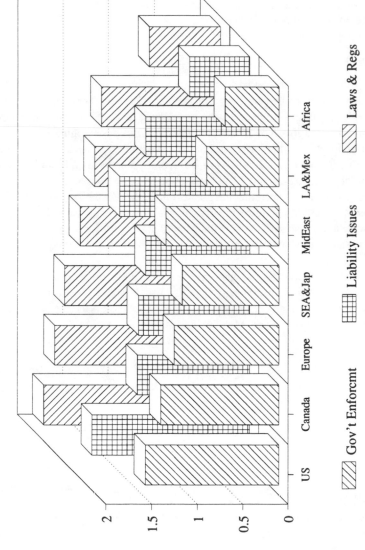

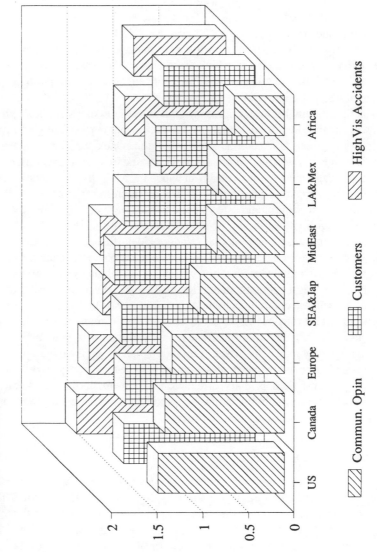

Figure 7.2
Influences on EHS Decisions

US Canada Europe SEA&Jap MidEast LA&Mex Africa

Commun. Opin   Customers   HighVis Accidents

procedures and protective technologies in place in all of their operations world-wide. One corporate environmental staff member observed: "If something bad happened, I would hate to have to explain that we do things differently outside the United States than inside."

As Chapter 3 indicates, some of the companies' policies specifically refer to the need for corporate standards where local regulations are nonexistent or where local regulations are sufficiently different that an effort to establish equivalency is appropriate. In such cases, MNC practice could exceed the requirements of host country regulations. One division EHS manager said: "In Spain, for example, I observed a situation in which there was an untreated discharge [from the facility] and I inquired of the local manager. He assured me that there were no air or water pollution laws in the country."

The expectation that legal regimes will become increasingly similar with respect to environment, health, and safety is one factor that is motivating companies to head in the direction of internal standardization. One approach to internal standardization is development of company-wide standards that are based on the most stringent requirements in any country in which the MNC has operations. However, none of the companies interviewed in this research has in place, for all aspects of environment, health, and safety, policies that were designed to meet the most stringent requirements in any of their countries of operation.

Those companies having internal standards identified selected areas of application; for example, one company had developed a company standard for a particular compound that is at least as stringent as the regulatory requirement of any country in which it operates. In another case, internal standards were developed in part to facilitate communication: "I introduced the OSHA standard for reporting accidents [throughout the division] because the Germans reported every finger cut, and in Spain, they only made a report when an arm was lost."

Although they may offer greater protection than the applicable regulations, internal standards cited by companies would still have to be considered "compliance driven" because their primary orientation is outward looking and anticipating government action, rather than inward looking and focusing on changing the internal corporate decision-making process.

CURRENT PRACTICE

In the case companies and other companies with which we spoke, practices were not uniform across all their operations. In discussing the issue of differences between home and host countries, one case company executive said: "Our marching order is to treat off-shore facilities the same as domestic. They are not equivalent, but we give [equivalent domestic standards] to them as the standard to achieve."

Several companies indicated that instead of uniformity, they are working toward "functional equivalence" in which procedures and equipment may vary across locations to accommodate local conditions, but the net effect to environment, health, and safety is the same from one operation to the next. A variation used in some companies is that of "equivalent risk."

However, interviews with case study environmental managers indicate that functional equivalence is not yet a reality in many operations. One case company EHS manager mentioned two locations, one in the United States and one in Germany, where their operating goal was "zero excursions at any price." He added: "In Brazil, we don't do a very good job. . . . However, the water going back out of the plant is better than it was coming in."

Corporate EHS staff from another case company mentioned a situation outside the United States in which "We could do more environmentally. We're not screwing things up, but if the economics change, we might do more." Another EHS manager said: "You can't export U.S. regulations or U.S. standards, and anyone who tells you that they are doing it isn't giving it to you straight." This view was reinforced by a corporate environmental staff person from a noncase company who said: "[This company] does not export 'U.S. or better' practices. A company has to go for equivalency. You can export equivalency, but you have to make value judgements."

All the case companies and several other companies interviewed in this research have operations in countries that lack facilities and infrastructure comparable to those in the industrialized nations, thus making functional equivalence difficult, if not impossible, to achieve in the short term. For example, a case company's Brazilian operation ships mercury waste off-site to an independent recycler. Given the state of enforcement in Brazil generally, it would be reasonable to predict that the recycler is not meeting standards that are functionally equivalent to those in the United States.

The absence of adequate hazardous waste disposal facilities was mentioned by all companies we interviewed that have operations in Brazil. One headquarters environmental manager said: "We have stuff piling up until there's a facility. . . . We met with an entrepreneur [in Brazil] and let him know what our requirements are." Clearly, stockpiling waste on the site of generation while waiting for a treatment and disposal facility to be built is not "functionally equivalent" to any practice in the United States.

In facility visits, researchers saw some practices and some facility design and maintenance features outside the United States that were less protective of the environment than those inside the United States. For example, at a facility in France, one of the process areas is located in a building that had previously been used for the manufacture of bricks. The floor is dirt, and process wastes seep in because the floor is permeable and impossible to clean. However, the converse was also true in that researchers saw examples of protective practices in facilities outside the United States that were not in place in the comparable U.S. facility. For example, a chlorine manufacturing facility in Brazil had implemented air monitoring at the fence line and a program of open gates to the community, neither of which were in place in the company's U.S. facility that researchers visited.

INTERNAL STANDARDIZATION AND SUSTAINABLE DEVELOPMENT

As in the case of international harmonization of EHS requirements, there are significant issues that need to be resolved for company-wide EHS standards to support sustainable development. In a world where nations increasingly compete for industrial investment, stringent headquarters environmental standards may exacerbate poorer nations' problems. Adhering slavishly to a program of global functional equivalence could, in theory, cause some investments to be passed over on the basis that local or national conditions make them too costly from an environmental standpoint.

The dual trends toward harmonization of EHS laws and harmonization of internal company standards will in some cases limit the options available to MNCs to deal with welfare tradeoff issues, and thus inadvertently work against sustainable development goals. Sustainable development is not current sacrifice for the development of a static world in the future; rather, it is a process in which "painful choices have to be made.[4]

NOTES

1. See Harris Gleckman, "Proposed Requirements for Transnational Corporations to Disclose Information on Product and Process Hazards," in *Corporate Disclosure of Environmental Risks: U.S. and European Law*, eds. Michael S. Baram and Daniel G. Partan (Salem, N.H.: Butterworth Legal Publishers, 1990), pp. 196, 197, for a full discussion of the "Home Country Legal Standard Approach."

2. T. H. Lafferre, Speech to American Institute of Chemical Engineers, Conference on Waste Minimization, Washington, D.C., December 4, 1989, p. 4.

3. See E. S. Woolard, "Corporate Environmentalism," Remarks by the chairman, Du Pont, before the American Chamber of Commerce, London, May 4, 1989.

4. World Commission on Environment and Development, *Our Common Future* (New York: Oxford University Press, 1987), p. 9.

8

Brazil and Mexico: Foreign Operations of *Oil and Gas* and *Household Products*

BRAZIL

Introduction

Brazil occupies the east-central region of South America. With a land mass greater than that of the continental United States, it is the fifth largest country in the world. Approximately one-half of Brazil is engulfed in dense jungle and rain forest. The magnitude and impressive nature of this landscape is often our primary identification with Latin America's most populous country.

Table 8.1 presents some comparative national statistics for the United States, Brazil, and Mexico.

The Brazilian economy is supported through rich natural resources of iron ore, manganese, bauxite, nickel, uranium, gemstones, and oil. A diversified agricultural base producing coffee, soybeans, sugarcane, cocoa, rice, beef, corn, oranges, cotton, and wheat reduce dependency on staple imports. Thirty-five percent of the active work force is engaged primarily in the industrial sectors representing steel, chemicals, petrochemicals, machinery, motor vehicles, consumer durables, cement, lumber, and shipbuilding.[1]

An increasingly urbanized population of 135 million people have settled or migrated to the industrialized areas of the south-central region. Four major groups make up the Brazilian population: the indigenous Indians; the Portuguese, who began colonizing in the 16th century; Africans brought to Brazil as slaves; and various European and oriental immigrant groups that have settled in Brazil since the mid-19th century. Although the basic ethnic stock of Brazil was once Portuguese, subsequent waves of immigration have contributed to a rich ethnic and cultural heritage.[2]

Table 8.1
Comparative National Statistics

COUNTRY	U.S.	BRAZIL	MEXICO
CO2 EMISSIONS FROM FOSSIL FUELS[a]	1135.3	41.5	73.7
HAZARDOUS AND SPECIAL WASTE[b]	250	33.9	21.9
PRIMARY ENERGY CONSUMPTION[c]	.61	.68	.56
# URBAN AREAS WITH 3MILL + INHABITANTS	6	3	1
POPULATION IN MILLIONS	241.6	138.5	79.6
AVERAGE LIFE EXPECTANCY	75	65	67.5
INDUSTRIAL ACTIVITY AS % OF GDP	27	30	35

([a] $= 10^6$ t a^{-1} as carbon; [b] = million metric tons per year; [c] = primary energy consumption per unit GDP 1984/1985.)

Source: World Resources Institute, *World Resources 1988/1989* (New York: Basic Books); *United Nations Environmental Data Report: 1989/90*, prepared for the United Nations Environmental Programme by the World Resources Institute and the Global Environmental Monitoring System (Oxford: Basil Blackwell Ltd.); World Statistics in Brief (New York: United Nations 1987); H. Jeffrey Leonard "Hazardous Wastes: The Crisis Spreads," *National Development*, April 1986, p. 33.

The vibrant and youthful society (50 percent of the population is under 20 years of age), is grappling with many socioeconomic conditions common to rapidly developing nations. An extremely narrow distribution of wealth exists, with the wealthiest 10 percent of the population accounting for over 50 percent of the income.[3]

Rapid deforestation in the north and lack of industrial planning in the south have exacerbated environmental degradation. Industrial and domestic sewerage is dumped directly into the many rivers, seriously impairing the public water supplies. Most cities lack the funds to collect sewage, and only a few of these treat it before releasing it into the rivers.[4] In greater São Paulo, only 5 percent of all sewage generated is treated. Automobiles are not required to have catalytic converters, nor are industrial flue gases adequately regulated. Sanitary control in the production of foodstuffs, even those produced for more sophisticated markets, is almost nonexistent: High doses of chemical preservatives whose use is prohibited or restricted in countries of the First World are widely used in Brazil.[5]

Brazil is divided administratively into 23 states, 3 territories, and the federal district of Brasilia. The framework of state and local governments closely parallels that of the federal government. Governors, elected for four-year terms, have far more limited powers than do their counterparts in the United States. This is due to the highly centralized nature of the Brazilian system and to a constitution that reserves to the central government all powers not specifically delegated to the states. The limited taxing authority granted to states and municipalities—the only territorial subdivisions of the states—has tended to weaken their power.[6]

Environment, Health, Safety: Regulatory Setting

For the regulation of EHS, Brazil has three levels of administrative capacity: federal, state, and municipal. The federal government's role has been to formulate a broad policy, leaving the more specific law and enforcement options to the state and municipal powers. The complex nature of this Brazilian constitutional approach has led to conflicting decisions and priorities, many times having detrimental effects on the environment. A further bureaucratic layer, created through federal legislation, is the formation of "metropolitan areas" such as São Paulo. "These do not constitute an autonomous level of government, only an administrative concept implemented by the states."[7]

From 1981 to 1983 the federal government enacted and applied the Brazilian Environment Act. This law calls for "the preservation, improvement and recuperation of the environmental quality, adequate for life, assuring the country of conditions for socio-economic development, for the interests of national security, and for the protection of the dignity of human life."[8]

Along with designating areas of responsibility for various institutions, this legislation establishes the means for implementing the environmental policy and presents important alterations in the approach to environmental issues, primaril᙮ by moving the regulatory center from pollution to prevention and protection.[9]

The state and municipal governments are empowered to exercise regulatory authority. The most influential of these powers is the ability to approve or reject operating licenses of any new projects that may have a significant environmental impact. "State and local levels can also impose fines, within the limits set by federal law."[10] Of the state agencies, the Environmental Sanitation Technology Company (CETESB) of São Paulo was the pioneer, establishing strict and comprehensive legislation in 1976 that deals with two phases of control on the subject, first on the preventive level and second on the corrective level.

In Brazil the three levels of government own or control many organizations chartered as private companies. CETESB was chartered as such, mainly to allow for a larger degree of administrative flexibility, especially in regard to personnel management. Chartered as a state-owned company in 1973, CETESB is the direct descendant of a line of state pollution control agencies, though its powers were initially restricted to water pollution matters.

Many of the Bazilians we interviewed believe regulated industry provided the impetus to create CETESB. These individuals describe a situation in which the water pollution had reached such severe levels that the industrial sector in the São Paulo region feared a shortage of water, even highly contaminated water, necessary to run their industrial processes.

CETESB provides expertise in technical and compliance activities. Specifically, this expertise is applied to installation, construction, expansion, and operation of any pollution source. This includes industrial, agricultural, and commercial establishments; equipment and machinery; or open-air combustion and self-propelled vehicles. Furthermore, CETESB has the ability to appraise projects and grant licenses, check business sites, impose fines and restrict activities, and shut down operations for up to 15 days. Revenues come from licensing fees, the rendering of technical services to other Brazilian institutions, and state grants. Currently, CETESB publishes the only comprehensive set of environmental legislation. The majority of this legislation, as well as operating standards, are modeled after EPA guidelines. As a CETESB spokesperson elaborated, "the standards are strict, but the analytical tools are just not available."

According to a government spokesperson, the major EHS problem facing Brazil is lack of low interest loans and financing mechanisms necessary for industries to make capital improvements in their pollution control technologies. Another significant problem for environmental, health, and safety regulators is a lack of "political will." According to a local regulator, "we need EHS issues to become a political priority in the government; this would lead to increased and improved enforcement of the existing legislation and we'd see immediate results."

As a branch of the Ministry of Labor, the secretariat of Occupational Safety and Medicine regulates safety and health conditions. These laws are presented in the Consolidation of Labor Laws (CLT, or Consolidacao des Leis do Trabalho). State and regional offices exercise varying degrees of enforcement activity. Representatives of the U.S. consulate in São Paulo discussed better health and safety conditions at MNCs due to management placing higher priority on labor satisfaction, and not the MNCs' ability to achieve technical superiority. Furthermore, local regulators cited nationality of the plant manager as an important factor in the EHS program, and they expressed a belief that U.S. managers have more effective programs. Stringent occupational health and safety laws are on the books; however, according to local regulators, there are inadequate resources for implementation and enforcement.

Federal government initiatives are currently under way in the following areas: (1) development of automobile emission standards; (2) implementation of automotive emission technology, that is, catalytic converters; (3) research into alternative fuel sources; (4) increased public awareness of environmental hazards; and (5) recent merger of the federal regulatory agency with the natural resource agency (IBAMA, the Brazilian Institute of Environment and Natural Resources).

The magnitude of impact that MNCs have on the business operations in Brazil is significant. In terms of employment alone, 154 foreign MNCs are among the

500 largest private companies (excluding banks and other financial companies) operating in the country, and they claimed gross sales of an estimated 45 perent of total national sales for 1987.[11]

Brazilian industry faces major challenges due to poor vocational and technical skills of the work force, a vast majority of whom are not functionally literate. For foreign MNCs, language and literacy barriers add additional challenges to training programs. Some companies are attempting emergency response drills at facilities that include surrounding communities, and according to a CETESB spokesperson, "we are working slowly and carefully in this area; the level of general public education is low and we fear a popular backlash against industry if we were to proceed too rapidly."

Overall, government priorities often conflict with long-term environmental, health, and safety issues. Providing basic nutrition, housing, water, and sewer service are often beyond the means of government coffers. Significant debt service, graft, and large-scale tax evasion compound dwindling financial resources.

FOREIGN OPERATION OF *OIL AND GAS*

History of Facility and Location

The facility we visited, Oilbras, is located in São Paulo state in the city of Cubatao. Cubatao is located near the coast, with the city of São Paulo to the north and Santos, Brazil's largest industrial port, to the southeast. São Paulo state is, by orders of magnitude, the most heavily industrialized region, although geographically it occupies only 2.9 percent of Brazil's total land area. The 11 million economically active Paulistas account for over 50 percent of all of Brazil's industrial revenue.[12]

Cubatao is located in what is referred to in the popular press as the "Valley of Death."

> A heavy industrial concentration has created a critical condition: the population suffers serious health problems as result of exposure to levels of pollution far above any acceptable standards. Operating in a relatively small area, twenty-three large industrial plants (plus dozens of smaller ones) are the major source of environmental degradation. These include a steel plant, a paper plant, a rubber plant, several nonmetallic manufacturing plants and assorted chemical plants. Five of these plants are federally owned and four are owned by multinational corporations.[13]

Oilbras is a 50/50 joint venture with Oil and Gas and a Brazilian partner. Oil and Gas has recently, within the past three years, acquired its portion of this partnership from another U.S.-based multinational. Oilbras has been in operation since 1964. According to the Brazilian HSE director, the other joint venture partner is not involved in management of environment, health, and safety. "They don't have the corporate structure or the philosophical approach to manage HSE issues."

Oil and Gas is currently involved in numerous business activities in Brazil, including but not limited to three wholly owned subsidiaries and three joint and limited partnerships. These activities are managed through a holding company, Oil and Gas of Brazil. As the company literature describes their activities, "Our permanent and active participation in the industry supplies the domestic market not only with chemical raw materials, but also with finished products of the highest quality as a result of constant application of advanced technology."

Products and Process

Oilbras is a supplier of chlorine and soda for the southern regions of Brazil. It provides chlorine to the market segments primarily involved in paper and cellulose production, chemical and petrochemical industries, water treatment, steel and metallurgy, aluminum and polyvinyl chloride (PVC) manufacturing, mining, and food processing. The approximately 600 employees run a 24-hour operation, with yearly production of 200,000 tons of chlorine, along with large quantities of liquid, flaked and solid soda, hydrochloric acid, and sodium hypochlorite. This facility utilizes both the diaphragm and the mercury cell electrolysis processes.

Environmental, Health, and Safety Issues

Waste management and disposal were discussed as the biggest problems for the facility to deal with. Other issues of concern involve the plant management's perception that environmental regulators lack technical expertise and preparedness when they conduct facility inspections.

Currently, there are no licensed hazardous waste landfills in Brazil. Some companies incinerate their wastes; many others are stockpiling, anticipating the construction of a disposal facility. The general talk among those we interviewed was of a lined landfill under construction in the state of Rio de Janeiro. Oilbras has developed markets for its waste products. For example, asbestos is being sold to a transite plant and mercury sludge is being sold to a processor who recovers the mercury and resells it to the plant.

There are some high noise areas in the plant that present a concern, and management is involved in examining some engineering alterations. In addition, the plant manager felt that in the heat of the summer months, the workers in the cell rooms suffer from varying degrees of heat stress. The plant is currently involved with the installation of an improved ventilation system. There are illumination problems, with over 50 percent of the facility falling below the lighting quality standard. This is undergoing an upgrade according to facility personnel.

Management discussed a program that involves monitoring health effects from mercury exposure. If elevated levels of mercury are detected in weekly urine samples, the individual is transferred to an area with less frequent or intense exposure. We were told this had no effect on an individual's salary.

The plant manager felt strongly that if technology existed that could "reasonably improve" an environmental, health, or safety related process, he would obtain it, regardless of import barriers. Furthermore, "Our first move is to protect the worker from a harmful exposure, then it goes into the air, from the air to the water, and each year we get progressively better in our improvements."

Environmental, Health, and Safety Policy

A spokesperson from the chemical industry group headquarters described Oilbras as:

> now, step by step, using Oil and Gas policies and figuring ways to implement them in Brazil. At our industrial site in Cubatao, São Paulo, two of the major concerns of Oilbras are safety and environmental protection. The company is proud of being considered by CETESB as a model of environmental protection, spending hundreds of thousands of dollars each year, protecting air, soil, water, and mainly human beings. With further regards to safety, various records have been achieved, reaching millions of man-hours worked without any injuries causing loss of work time.

Organizational Structure

The overall responsibility of this facility is with the industrial director (plant manager). He has three direct reports: administrative manager, operations manager, and technical manager.

The administrative manager is responsible for employee relations, health, and general services. The operations manager is responsible for production, maintenance, and quality control. The technical manager is responsible for engineering services, purchasing, and safety. The environmental coordinator reports up through the operations industry group.

The plant manager reports to the president of the joint venture, Oilbras, Inc. The president of Oilbras, Inc., in turn reports to the vice president of Oil and Gas of Brazil.

Management Philosophy and Style

Current management by a Brazilian national, as well as prior management by a U.S. national, were said to contribute to the strong EHS philosophy that exists at the plant. Much of the current emphasis on EHS was discussed as being a "holdover" from previous plant ownership. The individuals we interviewed stressed that they had been with the company through its acquisition and a few different managers. They spoke of the prior manager as having a profound impact on facility operation, particularly with his emphasis on environmental, health, and safety issues. "He made these issues a priority then and they are still the highest priority today."

The philosophy and culture of Oilbras claim strong autonomy from the Brazilian holding as well as from the parent company. Employees were proud of their work and the leadership role the company was taking in the region. Management spoke in terms of the "Oilbras Mission." There was a strong emphasis on teamwork.

The industrial site includes gardens, duck ponds, and other types of small wildlife. Management said, regarding the impact of these efforts on morale and efficiency of operations: "You couldn't pay Madison Avenue to develop a more sophisticated internal public relations campaign."

Furthermore, management practice is described as "highly centralized" within the facility. All decisions were seen as being funneled through the plant manager. Management is seen as "investing in its employees," meaning that many individuals are promoted from trainee levels up through skilled technician levels.

Environmental, Health, and Safety Programs

Oilbras is involved with the national chemical trade association, ABIQUIM, in developing a 24-hour hotline providing information on chemical issues. This is primarily for accident response either on transportation routes or at a facility. The hotline will expedite emergency response and provide toxicological information to rescue personnel.

Oilbras management describes itself as "playing an instrumental role" in installing the United Nations Environment Programme (UNEP) emergency response program APELL (Awareness and Preparedness for Emergencies at Local Level). They designed and carried out two practice drills leading to a seminar for all of South America for introduction of the APELL program. The seminar utilized a third drill including a mock community evacuation. Presenters and observers were present from local and national press (including national TV news coverage) and CETESB, and United Nations personnel attended from the UNEP Industry and Environment Office Paris headquarters.

A yearly "environmental week" is run in which local and state government leaders and regulatory authorities participate. One portion of this program is a dedication of a specific environmental project.

Plant management described a cost-benefit/feasibility study currently under way concerning the installation of a "Safer" system. This is an electronic system monitoring for chlorine; it electronically monitors wind direction, time, and temperature, and it plots a concentration plume to determine potential local neighborhood danger concentrations. "So far things look pretty feasible; now we just have to deal with import restrictions."

According to management the plant is worth approximately $180 million, and $16 million is invested in pollution control. The joint venture arrangement of the operation results in some restrictions for access to Brazilian financing.

Reporting Mechanisms

Industry group headquarters describes a program of an assessment once a year, with follow-up on the action plans that were developed during the previous visit. This occurs one time a year for environment, and another visit concentrates on safety issues. The action plans include recommendations and completion target dates. Each plant is requested to file a quarterly status report. These go initially to the Oil and Gas of Brazil environmental, health, and safety coordinator. He then moves them up to the chemical industry group technical center.

These assessments are characterized by the individuals conducting them as "not highly technical evaluations but to be used essentially to create a presence that has a quasi-regularity to it. We are in the process of developing a more systematic, regulation-oriented approach to address problems."

According to the plant manager, if a reportable incident in either the safety or environmental area occurs, an incident report is filed with Oil and Gas corporate headquarters within 24 hours. A detailed report is then requested within five days. These reports are also filed through the Brazilian HSE coordinator.

Approximately every three years, a four- to six-week EHS audit is conducted in conjunction with the normal financial audits. This audit covers compliance with local and Oil and Gas standards as well as action plan completion and procedures compliance.

Unions and HSE Issues

There is a union in the Cubatao valley that is primarily organized around the chemical and petroleum workers. A small percentage of the work force at Oilbras is unionized. These individuals tend to be in a more highly skilled labor market. The unionizing effort is gaining in numbers and credibility, and management is opposed to this. They feel they do not need the unions and that they have a good relationship with their workers. During a recent country-wide strike, attendance at the Oil and Gas facilities ran at 80 to 100 percent.

Community Relations

"Open Factory" is a program described by plant personnel as a 24-hour, around-the-clock program, that allows visitors into the facility to observe operations at any time of day or night. According to a company document,

In a pioneering attitude, Oilbras keeps its doors open to the community since November 1985. Through its Open Factory Program, any interested party may visit their installations, by means of a simple telephone request. Up to the end of 1987, the program resulted in an impressive number of approximately 4,000 visitors from a variety of segments, such as students, politicians, ecologists, clergymen and the public in general.

Plant management was enthusiastic about their lead role in the Cubatao area in developing community programs including an Environmental Week, an elementary school painting contest with an environmental theme, and an emergency response plan based on APELL.

MEXICO

Introduction

Experts predict the population of Mexico City will reach over 26 million by the year 2000, making it home to one in every five Mexicans. City officials attempt to manage the increasing population, but the daily influx of thousands of migrating campesinos from the countryside in search of work exacerbates the problems.

The lack of appropriate industrial and residential zoning is an endemic problem in the industrializing world. Mexico is no exception. The absence of urban planning, occurring with an enormous population explosion, has created a tight coexistence of heavy manufacturing operations within residential neighborhoods. The overall result has been a significant deterioration in environmental quality within the metropolitan region.

On a daily basis, Mexico City residents are faced with the exhaust discharged from over 3 million vehicles and the chemical discharges from over 36,000 industrial sites.[14] This heavy dose of 5 million tons of chemicals and suspended particles takes its toll on the health of the city population.[15]

The air quality issues have taken on political proportions, such that one corporate executive from a U.S. automobile manufacturer declared: "No Mexican politician is going to say that Americans have the right to breathe better air than the Mexican people."

The issue of air quality deterioration has created widespread public awareness while generating criticism of the government and its inability to address the growing crisis. The citizens of Mexico City pay close attention as the air quality is reported in a daily air quality index, which is published in the newspapers and announced on local radio stations. Many residents adjust their daily plans depending on the levels of pollution, avoiding outdoor activities when the index indicates particularly high levels of smog.

In terms of real risk, however, experts from the United Nations[16] feel that water pollution problems are the most immediate threat to the health and quality of life in Mexico City. An estimated 30 percent of the residents of the metropolitan area do not have sewage service and are forced to dispose of their wastes wherever and however they can.[17]

Regulatory Setting

The government agency handling environmental issues in Mexico is the Secretary of Urban Development and Ecology (SEDUE). It was referred to throughout interviews as "the Mexican EPA." SEDUE receives its authority from the ministry level of the federal government. The ministry is responsible for the "ecological policy" of the country, whereas SEDUE is responsible for the implementation and enforcement of a body of environmental law, originally instituted in 1972 and recently updated. The "policy" for environmental issues was described by one environmental regulator as "ambitious, but with a significant time lag between the policy and the technical capability." The Mexican environmental regulations are described as following an "EPA model, yet enforced with Mexican reality."

There are 31 states in addition to the Federal District (Mexico City) that constitute the United Mexican States. SEDUE has representatives located in each state charged with enforcing federal regulations, although state government is encouraged to toughen the federal laws when "appropriate." To date, five states have promulgated laws more stringent than the federal regulations, and 10 more states have entered a "serious negotiation" phase in regulatory upgrade.

SEDUE personnel are frequent participants in international training programs. Conferences coordinated in conjunction with the U.S. Environmental Protection Agency, the World Health Organization, the United Nations Environment Programme, and other environmental and health organizations provide training and curriculum updates in health effects, environmental control technologies, and regulatory developments in other countries in Latin America. Last year 250 SEDUE staff attended a training program in Texas sponsored by the EPA.

People interviewed for this research agreed that the present system of environmental regulation is not working. The regulations exist; the problem is lack of enforcement. A complete bureaucratic housecleaning takes place every four to five years following elections, resulting in no institutional memory from one administration to the next. Environmental regulators are technically undertrained, and financially underpaid, and tales of corruption run rampant. SEDUE has approximately 50 percent less staff than it did six years ago. The new government overall is managing on a 37 percent staff reduction.[18]

The major Mexican universities have begun, only within the past five to ten years, to graduate individuals with expertise in the environmental disciplines. At present, few qualified technocrats are available for employment in either the public or the private sector. The incidence of government workers taking bribes in exchange for ignoring pollution excursions was discussed during interviews as occurring fairly frequently. Many of those interviewed claim that the graft money is necessary for survival on a government employee's wages. Our interviews lead us to believe that the facilities of multinationals in Mexico are getting more consistent pressure for environmental improvement from headquarters or from sector bosses than from the Mexican government.

Health and safety issues are regulated by a variety of departments within the Ministry of Health and the Mexican Institute of Social Security. The Federal Labor Act provides the regulatory and enforcement authorization carried out by the Federal Labour Inspectorate.

The Institute of Social Security is supported in equal parts by the contributions of workers, their employers, and the government. An extensive system of hospitals and clinics is operated throughout the country to provide medical services to the general population, including the workers who may suffer from injuries or illnesses related to their occupation. Therefore, the Institute of Social Security has a major interest in reducing injuries and illnesses in the work place in order to control the costs for medical services and the payment of lost wages for injuries to workers.

The Institute offers a consultation service to employers to answer questions related to occupational safety and health. In addition, accidents are investigated and data compiled to assist employers in developing programs to aid employees in reducing injuries and illnesses in the workplace.

The various industrial enterprises in the country are divided into five groups according to the risk, injury, or illness to the workers in that classification, and classification determines employer payments to the Institute. The employers thus have an incentive to improve their safety and health programs in order to reduce the incidence of injury and illness and in turn receive a reduction in the contribution that must be paid for the support of the social security services.[19]

The organizational and bureaucratic apparatus of the Labour Inspectorate was described during our discussions as being more efficient than that of the SEDUE operations. However, SEDUE was characterized as more visible and influential in dealing with multinational corporations. It is possible that because the facilities we interviewed had relatively low incident rates for accidents, the frequency of interaction among health and safety professionals and the Labour Inspectorate was minimal.

Current initiatives and challenges faced by the government of Carlos Salinas de Gortari in the environmental, health, and safety arena include:

- A regional development plan that addresses industrial siting is continuing to expand. This program is aimed at encouraging, through favorable financing plans, the relocation of industry outside of metropolitan Mexico City.

- The government is requiring that all automobiles have catalytic converters by 1991, and it is working with the national petroleum industry, PEMEX (Petróleos Mexicanos) to upgrade its refining process to provide an improved fuel.

- Hazardous waste management is a looming problem for the government. This is being addressed by encouraging the privatization of waste disposal. A U.S.-based firm, Waste Management, Inc., has been given a contract to construct and operate a facility. They have been permitted to handle some solids, but the liquid waste treatment capability is still undergoing tests. A PCB incinerator was scheduled to undergo start-up operations in Tijuana in the fall of 1989.

- SEDUE has recently become more aggressive in its enforcement actions. This has resulted in closing down facilities in violation of the environmental laws, an infrequent and controversial maneuver for Mexican officials.

- In conjunction with the Finance Ministry, SEDUE assisted with the implementation of a resolution that provided significant tax rate reductions for industries making investments to improve environmental control technologies.

- The "maquiladoras" program of cross-border production sharing, although credited with creating jobs and foreign exchange, has also created a growing concern over illegal waste disposal and lack of infrastructure to accommodate vast population increases. A 1983 U.S.-Mexican agreement states that the Mexican plants can import raw materials and return products to the United States; however, if the process involves hazardous chemicals, they must be returned to the United States for disposal.[20]

NGOs and Trade Groups

The environmental movement in Mexico, although gaining momentum, wields relatively little clout among the government policymakers. "Not only are they (environmentalists) up against a political system designed to maintain the status quo but they also must battle political indifference to environmental issues as well as a general ignorance of the dangers of pollution."[21]

The NGOs use a censored media to raise issues in a public forum in an attempt to pressure the government into addressing the deterioration of air quality. The Grupo de los Cien, a loose alliance of 100 Latin American writers, artists, and intellectuals, has been one of the most vocal NGOs in public declarations against the government's environmental inactivity. For the time being, multinationals appear to be insulated from the criticism of the environmental alliances. These alliances are targeting the majority of their efforts at the government.

The Asociacion Nacional de la Industria Quimaca, A.C. (ANIQ), is a well-organized, highly visible trade association representing the chemical industry. Its membership includes 95 corporations, 12 of which are U.S.-based multinationals, including Household Products. Industry representatives spoke of ANIQ as a valuable source of information as well as a conduit between industry and SEDUE.

HOUSEHOLD PRODUCTS IN MEXICO

Overview

The Household Products manufacturing operation we visited, HPMex, is one of four facilities that the company owns and operates in Mexico City as wholly owned subsidiaries of the parent company. Household Products has owned this plant since the 1940s. The plant employs approximately 1,400 workers; 1,100 are hourly and 300 are professionally or technically trained. The plant operates on a 24-hour-per-day basis. The company environmental contact is charged with EHS responsibilities at all four of the Mexico City plants, rotating among the four facilities on a daily basis.

Products and Process

HPMex plant primarily manufactures detergents, which are further processed into household and industrial cleaning products. The process involves a mixing and distillation cycle, with raw materials on the receiving line and packaged consumer products on the shipping and distribution line. The primary waste product, spent acid, is combined with soda ash and shipped to the government-managed waste disposal facility in San Luis Potosi.

Environment, Health, and Safety Issues

The environmental contact, whose title is risk management coordinator, is a Mexican national fluent in English. His responsibility is "to assure compliance with both company policy and government policy on a cost-effective basis." Specific areas of responsibility include environmental controls on water and air emissions; process safety; fire protection; hazardous materials management; security; and industrial hygiene.

Additional EHS issues confronting the facility management include reduction of particulate emissions; waste disposal for both hazardous and solid waste materials; respiratory and eye ailments; lower back injuries caused by inefficient lifting and improper movements; and miscellaneous cuts and bruises.

The risk management coordinator sees his responsibilities as carried out across product division lines and as saving the company significant amounts of money. "It's difficult to prove the value of a job well done in the risk management area because if you're doing a good job, nobody hears anything about your activities."

The facility management recognizes that there are some environmental problems yet at the same time indicates a willingness to work with and accept the SEDUE regulators.

Environmental, Health, and Safety Policy

Maintaining a safe work environment is a well-developed and frequently articulated policy by the personnel at the HPMex plant. "If in doubt, safety first" is the overall facility theme. However, in discussions of environmental policies and programs, this enthusiasm and commitment is not matched. Staff articulated a commitment to improving the resources dedicated to environmental programs, yet, "the day-to-day safety functions take precedence over environmental management."

Management Philosophy and Style

Facility managers are committed to operating the Mexican plant in a self-sufficient manner. They are working toward developing the technical and management expertise in-house, relying less frequently on headquarters resources. Management communicated a cooperative relationship with competitors in the area of environmental problem solving.

Environmental, Health, and Safety Programs

Employees in the EHS department claimed that monthly safety committee and crew meetings have been successful in reducing the injury incidence rate and lost work time hours. Pressure and attention from corporate headquarters in the United States was said to have contributed to further reductions. A safety/industrial hygiene model that consists of values, programs, behavior influences, causes, and consequences, is an integral part of the program.

Some of the specific programs we discussed included promotions; workshops; behavior observation systems; positive recognition; enforcement; safe engineering design; analyzing performance problems; training; and chemical/technical review of hazardous systems.

The materials involved in the EHS programs were all written and refined at the facility level, with the guidelines provided from corporate headquarters in the United States. The start-up costs involved to initiate these programs were said to be more readily available than the resources necessary to improve and maintain them.

The core element in these programs is an approach to health and safety issues using a behavior modification strategy. They are designed to be directed at the hourly worker and supervisor levels. Recent emphasis is on "a more robust training for supervisors."

Reporting Mechanisms

The HPMex plant personnel use an Incident Reporting Form in communicating EHS issues both within the facility and to the parent company. This mechanism rates incident severity by placing valuations on four factors: environmental impact; public relations; regulatory involvement; and cost factors.

The facility has very explicit instructions in defining which events should be reported. For example, if any of the following criteria are met, the incident should be reported.

- event that harms human health (off-site) or the environment (on or off-site);
- event that results in noncompliance with environmental laws, regulations, permits, or company environmental policy;
- event that requires or results in notification or intervention of environmental authorities or results in penalties or fines;
- event that results in local, state, regional, or national media attention or public relations issues.

Community Relations

The majority of contact between facility personnel and the local community has been through the local police and fire brigade. The plant has run drills in an attempt to involve the local community in outreach efforts; however, the "people are just not interested in participating." HPMex has built children's playgrounds in the local community. Many members of the hourly work force live within close proximity to the facility.

NOTES

1. Rand McNally, *World Facts in Brief* (Chicago: Rand McNally & Company, 1986).

2. U.S. Department of State, Bureau of Public Affairs, Office of Public Communication, Editorial Division, *Background Notes, Brazil*, November 1987.

3. Oil and Gas company literature, May 1989.

4. Joao Carlos Pimenta, "Multinational Corporations and Industrial Pollution Control in São Paulo, Brazil," in *Multinational Corporations, Environment, and the Third World*, ed. Charles S. Pearson (Durham: Duke University Press, 1987).

5. Eduardo Viola, "The Ecologist Movement in Brazil, 1974–1986: From Environmentalism to Ecopolitics," *International Journal of Urban and Regional Research* 12 (1988).

6. U.S. Department of State, *Background Notes, Brazil*, p. 4.

7. Pimenta, "Multinational Corporations," p. 199.

8. Ibid., p. 201.

9. Ibid.

10. Ibid.

11. U.S. Department of State, *Background Notes, Brazil*, p. 11.

12. Pimenta, "Multinational Corporations," p. 206.

13. Ibid., p. 209.

14. Steve Nadis, "Mexican Clean-up," *Technoloay Review*, Nov./Dec. 1989, p.10.

15. Larry Rohter, "The 'Makesicko City,' New Smog Fear," *New York Times*, April 12, 1989, p. 4.

16. Information obtained during interviews at UNEP, Mexico City, August 1989.

17. Rohter, "The 'Makesicko City," p. 4.

18. Information obtained during interviews with Mexican environmental officials, August 1989, and Nadis, "Mexican Clean-up," p.10.

19. International Labour Organisation, "Safety and Health Practices of Multinational Enterprises," (Geneva: ILO, 1984), p. 22.

20. "Joint U.S., Mexican Manufacturing Program May Be Causing Pollution in Texas, Arizona," *International Environment Reporter*, June 1988, p. 306 and Jolie Solomon, "U.S. Firms' Standards in Mexico Targeted," *Boston Globe*, February 13, 1991, p. 31.

21. Jacqueline Moslo, "Commentary," *The News*, Mexico City, December 10, 1985.

9

What Drives a Good Program?

OVERVIEW

In this chapter we attempt to establish links between variables and derive at least one significant proposition that permits us to predict what drives good global EHS programs. In the process, each of the study propositions (first explained in Chapter 2) is examined based on the findings from the case companies and, where appropriate, modified propositions are offered.

We also examine the findings from our exploratory inquires on management policies and structures and standardization, and consider other factors contributing to good programs, such as evaluation tools and a long-term view in decision making.

SIGNIFICANCE OF PROPOSITIONS

The propositions advanced at the beginning of this book vary in their significance. For example, to conclude that bigger or more profitable companies or firms in certain industrial sectors can be predicted to have more effective EHS programs leaves little room for policy direction. These propositions have been included because previous research has identified them as potentially important variables. However, some of the propositions have greater significance because they are more closely related to phenomena over which corporate management and the public policy community have some control or influence.

UNITS OF ANALYSIS

For the purposes of this analysis, the EHS programs of companies are the units that will be used. These are not strictly the programs themselves as defined by the companies, that is, the structures and functions that appear on an organization chart or in job descriptions. The units as defined here are necessarily broader. They include not only the program components discussed earlier but also the management strategies, corporate policies, and overall shape and direction of corporate actions that together form a whole; have an effect on environment, health, and safety; and constitute each corporation's contribution to sustainable development.

EVALUATION CRITERIA

We build on previous work in formulating propositions for this work, and many of the propositions we use are causal statements linking a set of conditions and "good" or "effective" performance. To determine whether the findings from the cases support the propositions, "effective" has to be given a working definition.

Because there are no universally accepted measures of effective performance in the field, we selected our own set. It should be noted that had a different set of criteria been used (for example, numerical indicators such as number of regulatory violations, amount of lost work time, and ratio of EHS personnel to production staff), it is highly probable that the result would alter the ranking of the case companies. A recommendation resulting from our work is that additional research on performance measures be undertaken to facilitate both further research of this type and communication within and among companies about what works and what does not.

We chose the following criteria to compare companies and to test propositions related to program effectiveness: (1) consistency between the stated goals of headquarters and the statements and actions of facilities; (2) apparent ability to identify and solve EHS problems; (3) design and maintenance of facilities; (4) use of protective equipment at facilities; and (5) investments in EHS programs and people.

The evidence for evaluation included written materials from the companies and diverse other sources; information from regulatory agencies; direct observation at the facility; interviews with facility personnel; and interviews with a variety of corporate personnel as well as individuals from government, other companies, nongovernment organizations, and academia.

These criteria are essentially a series of discrete points from which to summarize the evidence and to formalize the subjective impressions of the researchers. The inevitable result of attempting to rank the five companies is that one company will occupy the number one slot in our terms of effectiveness, and another will be in the number five position. However, none of the case companies were uniformly worse than the others. Also, it should be remembered that this is a relative ranking of only the five case companies; we did not compare the five companies' performance to the universe of multinationals. The application of the criteria to the cases indicates where each company is in time, giving a snapshot of effectiveness. It will be understandably blurred given the speed at which each corporation is changing.

For some of the companies, researchers saw more facilities, talked with more people, and generally had greater access than others; no doubt this factor had some effect, but it is difficult to say whether it resulted in a more positive or a more negative assessment by researchers.

CASE COMPANIES IN THE CONTEXT
OF THE PROPOSITIONS

Proposition 1. There will be differences in the effectiveness of the companies' EHS practices that can be explained by the type of business in which each company is engaged.

The chemical operations of two of the case companies, Oil and Gas and Chemicals, were studied by researchers. If this proposition were supported by the cases, we would see similar programs with similar effectiveness in these two companies. Instead, there were differences in program components; according to the evaluation criteria, Oil and Gas had a better overall EHS program than Chemicals. For example, the Chemicals EHS operation in Europe was struggling for improvement in spite of corporate headquarters' intense focus on production and profitability. By the same token, EHS programs that were very similar in effectiveness were those of Instruments, Pulp and Paper, and Household Products, obviously very different businesses.

On the evidence, then, differences in industrial sector alone cannot reliably predict important differences in EHS program effectiveness. In other words, similarities in EHS program effectiveness need to be explained in terms other than industrial sector.

Proposition 2. Companies with greater consumer name recognition will have more protective environmental programs than companies with less name recognition.

Household Products has the highest consumer name recognition among the study companies. Household Products also has, according to our criteria, an EHS program that is relatively more effective than some in the study and much more effective than others. In interviews at headquarters, there was an acknowledgement that the name recognition gave the company a very strong business motivation to maintain high-quality programs.

On the other hand, both Pulp and Paper and Instruments are mounting relatively effective programs and have relatively less well-known names, especially among consumers. Their names, however, are very well known in their respective industries, within the finance and relevant business communities, and in the regions in which the companies have operations.

Chemicals, particularly in the divisions we studied, does not have significant consumer name recognition and also has the least effective EHS program of those studied. This finding tends to support the converse of Proposition 2.

On the evidence, there are some possible modifications of the proposition. It could be proposed instead that companies with good names to guard will have more protective environmental programs than companies without such names. For our purposes, however, the following is an even more persuasive revision: *Companies whose decision makers believe there are strong business reasons to be protective of the environment will be more motivated to have good programs than companies whose decision makers do not believe this.*

Proposition 3. The programs and decisions of small companies will be less protective of the environment than those of large companies.

This proposition is one that Roger Kasperson et al. observe is a common assertion with respect to hazard management,[1] yet we are aware of no comparative studies of performance in large and small firms. The assumption is based in part on the notion that firms with larger resources, even if they devote an equal percentage of those resources, will spend more in absolute terms than smaller companies. Usually, it is the very small companies that are implicated in this way, and the suggestion is that the cumulative impact of the actions of a large number of small companies is the real policy issue. Because this issue has received little rigorous study, no commonly accepted definitions of "large" and "small" have emerged. We chose to look at size from a relative point of view among the case companies, recognizing that other researchers might argue that no MNC could be characterized as being small.

Among the study companies, Instruments is a relatively small firm. Using 1989 as the reference year, Instruments had $504 million in sales and had 7,000 employees around the world, compared with the next smallest company in the study, which had $4.5 billion in sales and had 39,400 employees.[2] Yet Instruments' EHS program, according to the cited criteria, is on a par with the best programs in the study. Nor is effectiveness correlated to size among the four other case companies. According to this evidence, large size alone does not explain EHS program effectiveness, nor is small size a predictor of ineffectiveness.

However, the existence of even a relatively large number of small companies with good programs does not necessarily negate the general commonsense conclusion that large companies *have the capability* of establishing relatively better programs because they have relatively more resources to devote to problems. What remains is to explain with one or more of the other study propositions why Instruments has the effective program it does. The analysis of Proposition 1 suggests that it is not because of the business sector it is in. A partial explanation may lie in the fact that decision makers recognize that it makes good business sense to protect the company name through high-quality programs.

Proposition 4. Profitability of the corporation is important to strong EHS performance.

The expectation is that a profitable corporation is one that (1) has adequate resources to devote to EHS, and (2) is profitable because it is skilled at anticipating operating problems and good at government relations, including compliance.

Among the study companies, Instruments reported a $4.6 million net loss in 1989, largely a result of start-up costs associated with early versions of a new automation product. In rankings by *Forbes*, Instruments shows deficits in average return on equity for the last five years, whereas the most profitable study company had an average of more than 14 percent average return on equity over the same period.[3] Instruments also had, according to our criteria, an EHS program comparable in effectiveness to that of the most profitable company. On the basis of this case, Proposition 4 would have to be amended along the following lines: *Profitability is not as important to strong EHS performance as other factors.*

One area deserving further examination is the importance of profitability of a facility as opposed to profitability of the company as a whole in explaining EHS program differences. In companies in which we were granted access to multiple facilities in the same host countries, considerable variation in EHS effectiveness, as defined by the evaluation criteria, was observed. It is possible that in today's increasingly decentralized operations, a profitable parent corporation might have a facility whose EHS effectiveness could be low because of unprofitable conditions at the facility.

Proposition 5. Conformity with the EHS program developed by corporate headquarters diminishes as distance from headquarters increases and as cultural and political contexts become significantly different from those at headquarters.

This proposition was among the chief issues cited by many people after the Bhopal incident, and it is an area of concern to a number of companies. Note that lack of conformity with the corporate program need not mean that the facility is less protective of the environment and of health. Note also that conformity to the corporate program may not always be under the control of facility personnel. If the corporate policy is contrary to local customs, conformity creates an almost impossible situation for local managers. Finally, note that the proposition in its present form gives little credit to corporate policymakers who normally would be expected to recognize the cultural and political realities in program and policy development.

A test of this proposition among our study companies can be made by comparing adherence to headquarters policy between home and host country facilities. With respect to Pulp and Paper, the home country facility studied was in the same region as headquarters, and the non-U.S. facility was in Canada, in a regulatory and cultural setting more like headquarters than in any of the other cases.

The proposition as applied to this case company creates a problem. The home facility is in close contact with headquarters, operates very much in conformity with EHS directives, and manages a relatively high-quality EHS program according to the selected criteria. The Canadian facility is very much driven not only by federal and provincial law but also by the fact that it is a wholly owned Canadian subsidiary of Pulp and Paper. In addition, a strong cultural tradition of striving for independence from the United States and a very strong and militant union were also driving forces.

In this instance, the Canadian facility's EHS program, far from diminishing as distance from headquarters increases, is augmented by the various factors first mentioned. As noted in an earlier chapter, headquarters EHS people stated that the Canadian facility policy is actually an improvement on the headquarters document. The Canadian health and safety manager stated that in a number of specific program areas, his facility is significantly ahead of the U.S. facility. On the other hand, the strong headquarters drive for quality and continuous improvement as an overall management philosophy was thwarted in Canada on the facility floor, as union members felt threatened by power it gave to supervisory personnel.

In the case of Oil and Gas, the proposition needs to be modified at the outset in the recognition that headquarters sees itself as a holding company and, although corporate programs are developed there, the separate businesses function (and are encouraged to function) quite autonomously. With this caveat, researchers observed environmentally protective programs in place in the Oil and Gas facility in Brazil that were not in place in the stateside facility that produces the same product. This finding, along with the comparable finding from Pulp and Paper, raises some questions about the value of "conformity," when lack of conformity produces programs that are more progressive than those envisioned by headquarters and in place in the home country facilities.

The proposition in terms of Household Products has other problems. At the Mexican facility, very close conformity with company policy was the case, especially in terms of the health and safety program. The Deming model was very much in evidence; networking and team management were operating realities. This overall management process contained much of the thrust for quality and continuous improvement that is promoted from headquarters, and the EHS program as judged by our criteria was functioning relatively well. This was true despite very real cultural differences and a relatively weak government enforcement apparatus. The Household Products home facility, located in close proximity to the corporate headquarters, demonstrated fairly consistent environmental initiatives, whereas the health and safety program was inferior to that demonstrated at the Mexican plant.

Finally, the proposition needs considerable amendment in the case of Instruments. The stateside facility demonstrated close conformity with the EHS program as developed at headquarters. The EHS policy itself was considerably less formalized than in the other four cases, and there was no evidence of the Deming quality system; however, the programmatic components at the facility were not only in keeping with headquarters thinking but were effective according to the criteria developed for this assessment. In Instruments' European facility, despite some real cultural differences and a different regulatory situation, no significant diminution of the headquarters program was evident. Also, in terms of our criteria, the program was an effective one.

The fifth proposition, then, as seen through the cases, needs to be significantly amended. Evidence in a number of the cases suggests that as companies begin to grapple more seriously with overseas EHS compliance, less and less emphasis is placed on slavish adoption of specific procedures and more and more is placed on management systems and facility responsibility, allowing the scope to accommodate local regulatory and cultural differences and providing the opportunity for local innovation and adaptation. Thus, *program* diminishment (if a program can be said to consist of a number of concrete practices and procedures) will not be as serious a problem as an erosion of the *management system* caused by significant differences in cultural and political contexts. The problem here is that in some cases, management systems were in a state of flux as a function of downsizing and the introduction of new management approaches.

Given these challenges to the proposition, the following rephrasing may be more germane: Conformity with the EHS program developed by corporate headquarters can diminish as distance from headquarters increases and as cultural and political contexts become significantly different from those at headquarters. This need not have a negative outcome for health or environment nor need it increase corporate liability if an effective management system is in place that emphasizes quality and continuous improvement.

Proposition 6. The greater the top management commitment, the better the EHS performance.

This proposition has become axiomatic among a large number of interested corporate and government EHS advocates. Top management commitment can be measured in a number of ways, but usually includes a strong written policy signed by the CEO, some sort of board EHS committee or responsibility, and a senior vice president of EHS.

Among the study companies, top management commitment measured in these ways was present, although in variable degrees, in Oil and Gas, Pulp and Paper, and Household Products. Chemicals, although in the process of change, was not as advanced according to these measures of commitment. In terms of EHS effectiveness as judged by the evaluation criteria, neither Oil and Gas nor Chemicals had EHS programs that were as effective as those of Household Products and Pulp and Paper.

Instruments, whose EHS program was relatively effective as judged by the evaluation criteria, had none of these measures of top management commitment. It is possible that the company's small size in relation to the others (see Proposition 3) and relative informality create a management environment in which goals can become shared in the absence of formal actions from top management.

Looking more closely, Oil and Gas has a thinly staffed headquarters EHS capability, but much of the muscle and decision-making authority is at the division level in a very decentralized situation.

It is arguable in this case that top management also consists of the chief executives in the divisions, or subsidiary companies, and that our proposition can stand, provided we accept the fact that there is less commitment at these levels in the case of Oil and Gas. In Chemicals, the program was undergoing significant change, but the research did show less top management commitment and a less effective program.

In general, then, the case study research shows that there is a relationship between top management commitment and EHS effectiveness, provided the definition of top management is extended to include the top people in the divisions in decentralized situations. In other words, the relationship between top management commitment and EHS effectiveness is weakened in a decentralized situation unless division managers are also committed.

Proposition 7. Having a well-publicized environmental incident in the corporate history is a strong catalyst to development of protective EHS programs.

This is another belief that needs to be analyzed in terms of our case companies. Both Oil and Gas and Chemicals have histories of well-publicized environmental incidents. Among the cases, these two companies had the least effective EHS programs. Pulp and Paper and Instruments had no highly publicized incidents in their histories and by the same token had relatively effective EHS programs. Household Products has an effective EHS program according to the evaluation criteria, but no major environmental incident in its history.

Based on this evidence, having a well-publicized environmental incident in the corporate history may be a strong catalyst to development of protective EHS programs, but a number of other factors could be more important. These factors could be other causal variables raised in the present propositions, or they could be altogether new ones. It is probably true, for instance, that an environmental incident in *any* major corporation can represent a possible catalytic force in the upper echelons of other multinationals. For example, reactions of case companies and others to the incident in Bhopal suggest that it was the catalyst for assessing the extent of liabilities and, in some cases, for enhancing programs, not just in Union Carbide but in other multinationals as well.

CAUTIONARY NOTE

The foregoing brief analysis may serve as a cautionary note against accepting as gospel certain of the more common assertions about causal factors in effective EHS programs. It is possible, however, to consolidate and restate the initial propositions so that collectively they retain some meaning. One can say, for instance, that regardless of company size, industrial sector, consumer product recognition, profitability, or histories of well-publicized environmental incidents, EHS programs of MNCs are more likely to be effective when policies have the backing of top management, including division management; when management recognizes good business reasons for installing and implementing good programs; and when there are management systems in place that recognize the importance of cultural and political differences in overseas operations and are consistent with overall company quality goals.

This general proposition emphasizes the relative importance of management over other factors in determining EHS effectiveness and has the advantage of recognizing that program improvements can be wrought out of actions of corporations themselves and, in some cases, out of the actions of various public policy bodies. On the other hand, the proposition is rather severely limited in providing more detailed guidance for corporations who wish to control and limit EHS liability from overseas operations.

SUMMARY OF PROPOSITIONS

To answer the question, "What really drives effective EHS programs?" a number of commonly raised propositions were tested against the case companies. The analysis suggests that neither company size, industrial sector, nor profitability are necessarily crucial variables. It also suggests that energy put toward EHS programs by firms with a large consumer product base may in reality reflect a recognition that building a good reputation through effective EHS programs is only good business sense. The case research confirmed that conformity with headquarters programs can vary with distance and as cultural and political contexts differ, but that EHS practices need not suffer thereby, provided a management system is in place that recognizes the potential problem.

The analysis also suggests that commitment from top management is an important factor, with the recognition that in decentralized situations, top management should also include top people in the division and subsidiary. Furthermore, among the case companies there was no strong relationship between firms with a well-publicized environmental incident in their histories and current EHS effectiveness.

OTHER FINDINGS

Several of the findings from our exploratory research provide additional insight into what drives a good program.

Management Policies and Structures

Management structure and philosophy was found to have an influence on environment, health, and safety, as it often dictates which part of the organization must pay and who controls the unit that pays, both for environmental cleanup and for expertise. Organizational structures in the case companies had undergone recent change or were in the process of changing. The place on the organization chart for the corporate EHS group varied, as did the degree of centralization of the corporation.

The use of chargeback systems in two case companies distributes the costs of corporate EHS programs across the operating units according to the amount of corporate service used. This creates a situation in which a unit struggling with profitability may make a decision on how much to use corporate environmental resources based on ability to pay rather than on the environmental challenges it faces. In addition, chargeback was cited by case companies as creating confusion over whether corporate EHS has a consulting or a directive function. These issues are addressed in detail in Chapter 3.

Standardization

The case study MNCs do not have the same procedures and equipment in place for protecting environment, health, and safety in the operations in the United States as they have in those outside the United States. In some cases the non-U.S. operations appeared to be less protective, and in others they appeared to be more innovative than the comparable U.S. facility.

In cases in which there appeared to be less protection in facilities outside the United States, factors such as lack of enforcement by national or local government; inadequate infrastructure (communications, solid and hazardous waste disposal capacity); presence of a large temporary work force; difficulty in obtaining monitoring equipment; and competing priorities (concern for basic nutrition in the work force over concern for exposure to carcinogens) were cited by case companies as reasons for the differences observed.

It was the consensus of those we interviewed that having identical EHS programs in all of a corporation's operations around the world is not a realistic objective in the foreseeable future. The majority of case companies indicated that they are working toward worldwide functional equivalence, in which procedures and equipment may vary across locations to accommodate local conditions, but the net effect to environment, health, and safety will be the same in all operations.

Although many companies said they are aiming toward functional equivalence, equivalent systems are not a reality in countries where there is minimal infrastructure for waste management. One overall direction that could be pursued is the development of uniform international EHS standards. Such standards could benefit MNCs if the standards were perceived to be reasonable and enforcement were even. Currently, enforcement is uneven throughout the world, even within and among the industrialized countries. With balanced application of uniform standards, a company could feel secure that competitors were not benefiting by exploiting differences in environment, health, and safety regulations. We included a survey question on international practice to test views on this issue. Of the survey respondents, 76 percent agreed or strongly agreed that international standardization of EHS regulations would lead to improved practice. See Figure 9.1. We believe that this finding is important; future inquiries should confirm this result and also should examine the economic and trade implications of international standardization.

Figure 9.1
International Standardization of EHS Regulations Would Lead to Improved Practice

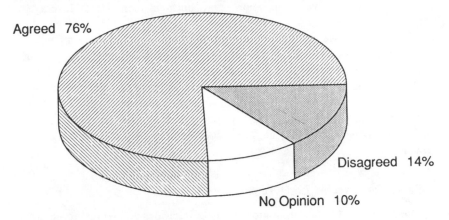

Agreed 76%

Disagreed 14%

No Opinion 10%

Information and Expertise

Considering the resources that are theoretically available to managers in MNCs, we found that EHS managers relied on a relatively small number of sources of information and expertise in developing and implementing programs. In the companies we interviewed, EHS managers rely extensively on informal networks of corporate peers for information on EHS programs. Other sources of information and expertise used in case companies include environmental groups, trade associations, consultants, governments, and internal experts.

Systems for Reporting and Evaluation

Accountability systems for environment, health, and safety within corporations are changing and are becoming more formalized while other aspects of corporate operations are becoming less hierarchical; some case companies used ad hoc teams to manage several aspects of environment, health, and safety. Case companies indicated that they use different systems for internal communication; however, regular reporting of a variety of indicators related to environment, health, and safety is being implemented in four of the five case companies. Application of these systems in non-U.S. operations lags behind application in U.S. operations in all cases.

One challenge faced by all case companies is determining what indicators of EHS performance to use. One company has settled on violations of government regulations and excursions from corporate standards, recognizing that there is variability in regulations and individual interpretation of standards across its operations.

Audits or assessments of various types were identified by case companies as valuable managerial tools. Baseline environmental surveys at facilities were not generally performed, and audits were primarily compliance-oriented. Four of the case companies indicated that audits were increasing in formality and frequency; however, there was variation among companies in their use as leverage for improved practice.

There was also variation among companies with respect to the composition of audit teams and dissemination of audit results within the company. For example, a case company stated that in one instance in which unacceptable practice was identified through an audit, no written record was made, and the matter was addressed with telephone calls to top management. This type of action reflects the fact that in the United States, liability concerns drive many corporate EHS actions, and lawyers often play an important role in helping to establish company practice.

Short Term versus Long Term

The tendency of U.S. corporations to make decisions that will increase profitability in the short term has a negative effect on environment, health, and safety. Case companies cited several specifics, including the following: (1) Investments that will benefit environment, health, and safety (as well as productivity in many cases) may take longer to pay off than the two-year average for most businesses within case companies; and (2) Development of technological innovations in both product and process is a time-consuming and costly process and is difficult to justify if the primary motivation is environmental. Consequently, companies are responding to outside pressure from regulators and the public rather than developing a long-term strategy for environment.

A question in our survey lends an interesting perspective to this issue. Respondents were provided with a list of eight factors and asked to select the three most important ones preventing their companies from doing a better job at EHS. Focus on short-term profitability was most frequently cited as the reason, with 53 respondents choosing it as a factor. See Figure 9.2. Limits of technology was the factor selected least frequently, indicating that respondents perceive institutional and managerial barriers to be greater than technological barriers.

This finding is important because it suggests that improved EHS will come through eliminating the institutional and managerial barriers to use of existing technology rather than from the development of better technology. Playing this out, it is our sense that a vice president for environment, health, and safety with a concern for the issues and with strong managerial skills will push a company ahead farther and faster than will a top-notch scientist, lawyer, or technician who has little experience in management. We see motivating people, communicating effectively, establishing influence through networks, and creating mechanisms for changing corporate culture as being absolutely critical to making significant progress in corporate environmental management in the years ahead.

Figure 9.2
Factors Preventing Better EHS

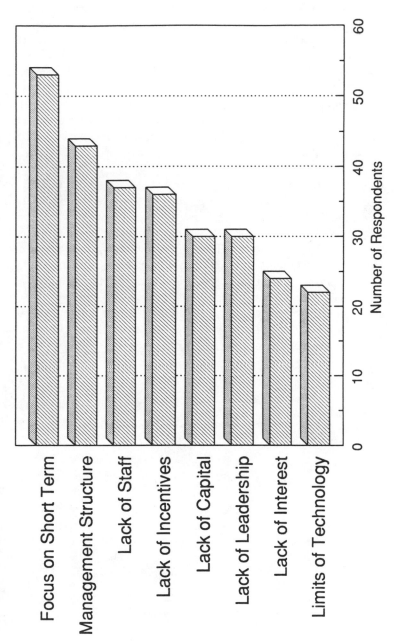

Number of Respondents

Acquisitions and Divestitures

An activity having implications for environment, health, and safety in all case companies is acquisitions and divestitures. In planning this study, researchers did no identify acquisitions and divestitures as having an important influence; however, we found that significant resources were invested by case companies in these activities. Again, this is an area in which corporate action is driven by concern for liability.

When both assets and whole businesses are acquired, EHS personnel often act as resources to business decision makers regarding the nature of potential problems associated with land contamination and hazardous processes. EHS personnel indicated that their findings can have an impact on price in business negotiations.

When businesses are acquired, issues related to compatibility of EHS programs with those of the new corporate parent must be addressed by headquarters staff. The cases reported here suggest that acquisition of personnel as part of a business purchase may have a stronger influence on the quality of the program in place in the facility than the program directives from the new parent's corporate headquarters. This is an area in which further research is needed.

The survey data on this issue sheds an interesting light on how a broader group of companies view this issue. Respondents were asked how often environment, health, and safety issues are addressed in decisions to acquire new businesses. The results, illustrated in Figure 9.3, are of particular interest because they suggest that for the majority of companies, EHS is not a factor in strategic business decisions.

Figure 9.3
How Often Are Environment, Health, and Safety Issues Addressed in Decisions to Acquire New Businesses?

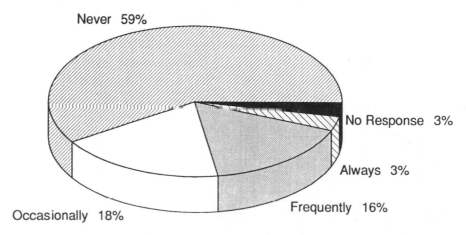

One corporate EHS manager, when asked how he was involved in the acquisition process, said; "Things move so quickly around here, we're usually brought in after the financial and business decisions are made." The timing of EHS involvement may be an important factor. A September 1990 survey by Deloitte & Touche and Stanford University found that 91 percent of firms surveyed evaluate the environmental performance of potential partners.[4] Our finding is not necessarily contradictory, and it raises the possibility of a subtle distinction: that in most cases the EHS evaluation occurs, but *after* the strategic business decision is made.

The foregoing discussion has a strong process orientation, that is, a focus on how decisions are made and what factors are most influential in corporate actions that affect the environment. An equally valid approach, but one that is considerably more challenging for the researcher, is to examine the outcome of these decisions as it relates to sustainable development. A worldwide movement toward sustainable development has the possibility of being integrated within corporations, though perhaps not by that name. Sustainable development could become an additional element in a determination of factors that govern MNC environmental practice. Aspects of the relationship between sustainable development and the results of MNC management innovation are discussed further in the next chapter.

NOTES

1. Roger Kasperson et al., *Corporate Management of Health and Safety Hazards: A Comparison of Current Practice* (Boulder: Westview Press, 1988), p. 127.
2. Dun's Marketing Service, *America's Corporate Families,* vols. 1 and 2 (Parsippany, N.J.: Dun and Bradstreet Co., 1989).
3. "Who's Where in Profitability," *Forbes*, January 9, 1989.
4. Deloitte & Touche and Stanford University Graduate School of Business, Public Management Program, "The Environmental Transformation of U.S. Industry: A Survey of U.S. Industrial Corporations, Environmental Strategies, Management Policies and Perceptions" (1990), p. 9.

10

Multinationals and the Environment: The Challenges of Sustainable Development

INTRODUCTION

The first part of this chapter examines some of the means by which corporate activities can be reinforced to promote sustainable development. Various definitions of sustainable development and their relevance to corporate EHS management are presented, followed by a discussion of the trends toward global EHS standards and corporate proactive initiatives.

Against this background, the role of the corporate EHS manager and management system in sustainable development is analyzed. Limitations placed on corporate initiatives in support of sustainable development are discussed, as is the need for flexibility in response. The role other institutions can play in supporting greater corporate initiatives toward sustainable development is explored.

The second part of the chapter presents our conclusions and recommendations on global corporate environment, health, and safety programs.

SUSTAINABLE DEVELOPMENT

The World Commission on Environment and Development defined *sustainable development* as a process of change in which the exploitation of resources, direction of investments, the orientation of technological development, and the objectives of institutions are all in harmony and enhance both the current and future potential to meet human needs and aspirations.[1]

The report of the World Commission was debated and accepted by the General Assembly of the United Nations. Although the term *sustainable development* has appeared in the literature since the mid-1970s, the World Commission popularized its use. Various actors in the environment/development community do, however, have reservations about sustainable development. In particular, a number of economic development institutions have noted that the concept has yet to be proved applicable at the project level, where concrete decisions are made.

To the extent that individuals from MNCs use the term at all, the majority interviewed in this research used it as a shorthand for economic development with environmental protection. For example, the headquarters staff from one of the case companies observed of countries in the Pacific Rim: "These countries have differences in technical know-how and in philosophy. They are looking for sustainable resources that they can develop."

In another company, there was a strong indication that the concept was not part of top management's thinking in 1989:

> Q: How do you get the company's attention on the issue of sustainable development?
> A: There is a major issue in getting attention for anything that is not directly related to the business portion of the business. When CEOs step back and look at the big issues, they're dealing with things like trade barriers and the implications of 1992 in Europe, and so forth. [The chairman] has probably never heard of sustainable development. And it is not clear it could be framed in terms that he could work with.

It is important to emphasize, however, that the exchange just quoted might be quite different if it were conducted today. Concerted efforts on the part of industry organizations, such as the International Chamber of Commerce, have called attention to the role of corporations in making sustainable development a reality.

Since the notion of sustainable development has proved to have utility in creating alliances between pure environmentalists and pure economic development advocates, it has gained some general acceptance, although further work on developing an operational definition will clearly be useful in directing strategic and policy discussions.

In *Our Common Future*, the World Commission on Environment and Development lists a number of general conditions that nations should have in place to advance sustainable development. These requirements, a summary of which follows, are not in place in many nations.

- a political system that secures effective citizen participation in decision making;
- an economic system that is able to generate surpluses and technical knowledge on a self-reliant and sustained basis;
- a social system that provides for solutions for the tensions arising from disharmonious development;
- a production system that respects the obligation to preserve the ecological base for development;
- a technological system that can search continuously for new solutions;

- an international system that fosters sustainable patterns of trade and finance;
- an administrative system that is flexible and has the capacity for self-correction.[2]

In addition, the Commission identified several specific actions that governments generally should be expected to undertake, including the following:

- the establishment of environmental goals, regulations, incentives, and standards;
- a more effective use of economic instruments that encourage industries to internalize environmental costs and reflect them in the prices of products, such as in the 1972 OECD Polluter Pays Principle (PPP);
- broadening environmental assessments to include not only products and projects but also policies and programs, "especially major macroeconomic, finance and sectoral policies that induce significant impacts on the environment."[3]

The Commission's list for industry action includes the following:

- accepting a broad sense of social responsibility;
- ensuring environmental awareness at all levels;
- establishing company-wide policies regarding resource use and environmental management, "including compliance with the laws and requirements of the country in which they operate."[4]

In addition, the Commission believes that international trade associations "should establish and make widely available sectoral guidelines for assessing the sustainability and potential hazards of new facilities, for developing accident contingency plans and for selecting pollution control or waste treatment technologies.[5]

The UNCTC has prepared a report that contains criteria for sustainable development management. According to UNCTC, "These criteria seek to provide a foundation for a new corporate management approach to environment and development,"[6] and they specify concrete corporate first steps toward "environmentally sustainable development." Topics addressed by the criteria are as follows:

Time and Space
    Time horizons
    Spatial boundaries
Living with Nature
    Stewardship of common resources
    Managing uncertainties
Efficient Technologies
    Efficiency of production
    Sustainable technology
Diversity
    Diversification of production
    Bio-diversity and socio-diversity

Global Consequences
    Reallocation of activities
    Atmospheric changes
Auditing and Reporting
    Auditing and assessment of performance
    Reporting successes and failures
Continuing Education
    Anticipatory thinking
    Participatory learning[7]

Although some MNCs have publicly subscribed to sustainable development goals, none to our knowledge, including those in this study, have comprehensively undertaken the specific steps called for in the UNCTC document.

In interviews and conversations with EHS managers, reactions to the concept of sustainable development ranged from no knowledge of the concept at all to a general endorsement of the idea, as it implies a rejection of the extreme environmentalism and no-growth animosities that characterized much of the rhetoric of the last two decades. Some managers believe that energy efficiency is the key to sustainable development and have no quarrel with the term in that sense, but they argued that it had very little to do with the risk-based compliance activities for which they were responsible. As one case company executive observed, "We're not trying to save the whole world; our job is to keep the company doing things right."

COMPLIANCE PLUS—OR MINUS?

In a number of interviews and in the public and policy statements of many U.S. MNCs, it is common to hear that the company "goes beyond compliance" and is "proactive." This has a very specific meaning in the United States, where companies are increasingly attempting to anticipate more stringent laws and more strident public opinion.

Internationally, the situation is more ambiguous. Going beyond compliance in environment, health, and safety may or may not address the substantive differences in each host country's national welfare and culture. More often than not, it can reflect a struggle by the company to stay ahead of corporate liability and responsibility in the context of an international legal process that fails to address variations in local resources and local ability to manage environmental issues. In this study, companies striving for maximum protection against strict legal liability in Brazil and Mexico may have been working against the goals of sustainable development.

For instance, two of the study companies initiated programs of community training and emergency preparedness in their respective host countries based on a U.S. model. They initiated these programs for a variety of reasons, including anticipation of local legislation and a desire to decrease the risk of disaster in the event of an accident at their operations. In both cases, although local officials and emergency response agencies were enthusiastic, participation and interest on the part of the local population was poor. Perhaps there is a need for flexible approaches that seek "to discover the ways and incentives characteristic of the culture that are more likely to result in acceptance."[8] Such approaches would also be more consistent with the goals of sustainable development as outlined by the World Commission on Environment and Development.

As an alternative, each of the two companies could, for example, prepare for evacuation of the population by assisting in the improvement of local road and transportation facilities. For instance, there is only one major road leading out of an industrial town where a case company's facility is located. Similar actions, such as the renovation of local and regional clinics, can minimize potential damage associated with a major industrial accident while simultaneously meeting some of the goals of sustainable development.

It is important to note, however, that the goals of sustainable development and the establishment of uniform EHS standards that are goal-oriented are not necessarily in conflict. To the extent that they complement one another, a significant burden of responsibility is placed on headquarters managers to mount and enforce facility actions that fully reflect an enlightened corporate policy.

TOWARD IMPROVED INTERNAL EHS MANAGEMENT

It cannot be emphasized too strongly that the global upsurge of EHS activity has meant that the EHS management systems of virtually all MNCs are in the process of change. This has certainly been the case with the study companies. It follows that no study company can be said to have either completely well-developed procedures or completely successful implementation. In some cases, the contrast between corporate policy and facility execution has been fairly sharp. This does not mean that the facility procedures are worse than similar operations in the same host countries.

LIMITATIONS OF HEADQUARTERS EHS MANAGEMENT

EHS managers must often work against powerful institutional and cultural constraints and against a background of shifting societal goals. Moreover, many of the central policy decisions that could make a meaningful contribution to sustainable development are often not in the hands of the headquarters EHS managers. For example, a company's decision to open up a major coal mining joint venture in China, despite the obvious implications for global carbon dioxide emissions, was made as a high-level business decision in which environmental personnel did not participate. EHS management was called in after the fact to address the pollution control and health and safety aspects of the operation.

EHS managers also have had little or no influence over the choice of new sites for manufacturing; the product to be manufactured; the process to use; the production schedule to follow; the past and present actions of neighboring local, state, or private firms; or the level of pollution of upstream surface or groundwater. In addition, the technical sophistication and capabilities of subcontractors and government regulators are variable and not easily improved. Above all, EHS managers have had very little influence over the values and motives of the dominant corporate culture that controls the priorities of the division and facility managers.

CORPORATE INITIATIVES TO PROMOTE SUSTAINABLE DEVELOPMENT

Some companies have developed EHS programs in which there is enough flexibility for managers to take actions that promote sustainable development goals. Many companies, however, approach EHS management from an explicitly home country, corporate, and short-term point of view. Although the UNCTC has provided guidelines for EHS management in support of sustainable development,[9] there has been relatively little specific treatment of sustainable development by the sources of information and expertise that companies indicate they find useful, including consultants and industry organizations. The exception is the International Chamber of Commerce's "Business Charter for Sustainable Development," mentioned in Chapter 1.

In the case companies, actions that can be characterized as supporting sustainable development were taken on an ad hoc basis rather than as part of an overall company plan. In Brazil, for example, a case company found that its high-pressure gas line had been tapped and was being used for cooking in the nearby community. Because the gas was not odorized, leaks could not be detected and therefore posed a significant hazard to the illicit users. The company eventually ran an odorized low-pressure line into the area to provide free access to the gas by the local community. According to headquarters EHS staff, "This is a case of something you work out."

In another case company, the former manager of a joint venture in the People's Republic of China described a circumstance in which he arrived at the facility and found all of the workers barefoot. It took the company two months to determine that sufficient shoes were available and that they could obtain steel-toed shoes for the machine shop. According to the former manager, "As part of its social welfare program, the company bought the shoes." The joint venture manager did not even clear this decision with his boss, indicating that he did not see any other way to deal with the situation: "It was not debatable."

Two issues that EHS programs consistently fail to explicitly address are the multicultural context of managing environment, health, and safety in MNCs, and the need for greater diffusion of responsibility on EHS matters outside the EHS management hierarchy. Both of these issues are intimately related to the degree to which EHS management is centralized and communicated. Efficient and effective global EHS management requires a systems approach that permits a maximum of autonomy at the facility level and provides adequate "ownership" of environmental problems at the factory floor where everything finally comes together. Making these systems work can be difficult even in home country settings. For example, one of our study companies has a group of facilities within close proximity to headquarters. When asked whether he takes corporate policy seriously, a facility environmental manager said: "Of course, whenever it makes good business sense."

Recent trends toward decentralization and greater facility autonomy complicate the control problem. Language barriers can exacerbate it even further. Both Oil and Gas and Household Products in Mexico are entirely run by Mexican nationals. This is a major improvement over the earlier days, in that the new managers are more closely in tune with the cultural environment (and they also cost the company less money), but it puts a burden on corporate and division management to communicate and to work together effectively. These institutional barriers are not trivial. In one foreign facility, a plant manager resolutely refused to recognize the authority of his headquarters superior. In another instance, a very senior headquarters EHS vice president from a U.S. chemical corporation (not a case study company) was not permitted to visit one of the company's facilities, which was a wholly owned Brazilian subsidiary. He was told there were union problems and that the unions would never let him into the plant.

Several measures can be taken to ensure control while encouraging greater facility autonomy. One means is through a more comprehensive and longer-term corporate EHS perspective than one resulting from compliance-driven activities. To be effective, this means that overall EHS goals need to be extended and broadened throughout the company and absorbed into the culture. Environment, health, and safety become a goal for the entire corporation. Such a prescription is consistent with the view that Kenichi Ohmae paints of the successful global corporation: "[C]ompanies must denationalize their operations and create a system of values shared by company managers around the globe to replace the glue a nation-based orientation once provided. The best organizations operate in this fashion and, as a result, devote much of their 'corporate' attention to defining personnel systems and the like that are country neutral."[10] In such a context, opportunity exists for corporate and plant EHS managers to become central actors in this evolution, thus enhancing their own careers and becoming key actors in the overall corporate contribution to sustainable development.

In some countries, novel or unique approaches to EHS management may be necessary to achieve levels of protection of the local population and ecosystems that are consistent with sustainable development. There needs to be some flexibility in the communication process and local accommodation. One means of providing local or national flexibility is through the inclusion of a formalized cultural impact procedure. Such a process could explore aspects of the local culture that are highly relevant to EHS management, such as beliefs and values toward health, safety, and risk; beliefs and values toward the natural environment; and the perceived relative value of the facility to the community. Managers could include discussions with previously established corporations as part of the cultural impact procedure. If the goal of the corporation is a relatively restricted range of risk and outcome for all of its facilities, then there is a need for flexibility at the facility level to provide the most efficient mix of tools to accomplish the goals of sustainable development.

EXTERNAL CHANGES SUPPORTING
SUSTAINABLE DEVELOPMENT

This section briefly addresses actions by governments and universities that will enhance corporate environmental efforts. It is important to emphasize that many of the points made in the following sections are environmental examples of broader recommendations made by the Massachusetts Institute of Technology's Commission on Industrial Productivity.[11] This is clear evidence of the passing of both the old "industry versus environment" mentality and the recognition that all sectors must cooperate if important societal goals are to be met.

Government Actions

Government's continued exercise of regulatory and economic controls is important to continued progress toward sustainable development; however, government must create a climate of certainty in the regulatory process that has often been absent in actions to date. The conflicts associated with environmental protection versus economic growth continue to hamper the regulatory process.

One argument used against government developing more stringent regulations is that productivity may be impaired due to the increased costs of management oversight and technological modifications that accompany regulations. Our survey results shed an interesting light on this subject, as Figure 10.1 illustrates. In response to the statement, "Stronger environmental regulations cause decreased productivity," fifty-five percent disagreed or strongly disagreed with the statement and 19 percent agreed or strongly agreed, thus indicating that the majority of respondents do not support the assertion that stronger regulations decrease productivity.

In the companies we studied, several environmental managers identified a need for an extension of the strategic business time frame in order to increase investments in environmentally beneficial projects. At its simplest level, the corporate perception of long term versus short term is critical in determining the kind of payback that decision makers are willing to consider and, therefore, the kind of EHS investments that can be "justified." In practice, however, the importance of a planning horizon for EHS is more profound. As one EHS manager observed, "What I want to do is get into a competitive position with respect to anticipating EHS regulations. If you anticipate, you can develop technology and solutions on your schedule. . . . Five years is a long time—very long—we're doing well if we can get people to think ahead three to four years. The normal time frame for making business decisions varies from quarterly to one year."

Figure 10.1
Stronger Environmental Regulations Cause Decreased Productivity

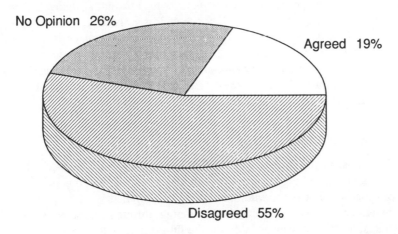

No Opinion 26%

Agreed 19%

Disagreed 55%

The role of the government in driving corporate EHS time frames is important. In the United States, the market for environmental protection technology is defined by EPA regulations, and those regulations have changed with regularity, but not with predictability. Sheldon Novick et al. found that the congressional attempt to create "technology forcing" amendments to the Clean Air Act in 1970 failed because the legislative time frames were too short to permit development of technology,[12] a view that is reinforced in a broader study by ICF.[13]

The advantage of long-term thinking, particularly with respect to the environment, is that it provides the ability to integrate multiple aspects of corporate resources toward achievement of goals. For example, people with skills in research, product development, purchasing, production, and so forth can be encouraged to focus, each from his or her own perspective, on achieving environmental objectives. Furthermore, if companies were to take a longer-term view, retrofitting and restructuring costs resulting from increasingly stringent environmental regulations would be reduced.

Government can encourage development and transfer of environmental protection technology. The U.S. Federal Technology Transfer Act of 1986[14] is an example of an effort with broad motivations that may have specific benefits in environment, health, and safety. The act and accompanying executive office actions have changed patent policy so that firms of any size making inventions may now own "title to patents made in whole or in part with Federal funds, in exchange for royalty-free use by or on behalf of the government."[15]

In the past, only small firms were permitted to retain patents on work with government funds, thereby creating a significant disincentive to large firms' investment in development. It is too early to tell whether this change will have a dramatic impact on environmental technology; however, it represents progress in eliminating barriers and provides a model for other countries to examine barriers to development of environmental technology.

Another type of government action that can lead toward sustainable development is a direct request to an MNC for technical assistance. For example, Oil and Gas was asked by a host government to assist in fresh-water exploration activities. The company had the technical resources to respond to the request and regarded it as a condition of doing business in the host country. Such an approach, if applied systematically, could create the basis for effective use of MNC skills in solving host country environment and development problems.

University Programs

Educational institutions in general, but universities and graduate schools of business and engineering in particular, can play a key role in promoting corporate actions that are consistent with sustainable development.

Universities can assist students in developing, through course work and other programs, a sense of responsibility for the future. Examples from the case companies have shown that the current emphasis on short-term profitability in the United States has precluded many investments in EHS improvement. If all actors in the system, from stockholders to consumers to individual corporate decision makers, were educated to think about long-term as well as short-term gains, an important redefinition of what is "in the corporation's best interest" would occur.

This objective can be accomplished with a specific "environmental literacy" program; however, it is possible to envision other emphases, such as social responsibility, that might have comparable benefits in encouraging individual action in support of sustainable development. The Tufts Environmental Literacy Institute, initiated in 1990 with funding from Allied-Signal Corporation, is one approach. This program develops the capacity of existing faculty in any discipline to modify courses to include an environmental element.[16]

Educational institutions can place priority on integrating environmental considerations into the curriculum. The environmental field offers a rich set of examples that can be drawn upon by a variety of disciplines. For example, in a course on international economics, debt for nature swaps and their impact on national policy could be discussed. Similarly, in a course on corporate strategy, students could discuss Du Pont's decision to stop manufacturing CFCs over a 10-year period because the compounds have been linked to ozone layer destruction.[17]

Two areas in which environmental integration might be particularly valuable to corporations are engineering education and management school curricula.[18] If environment is considered by engineers from the very beginning stages of product development, issues such as disposal of worn-out products and empty packages will be addressed before products are even brought to market.[19] Graduates of management schools are important to enhancing the corporate response to EHS challenges, because they develop expertise in motivating people and guiding organizations to achieve objectives. To the degree that these objectives are viewed as consistent with sustainable development, both corporations and society will benefit.

Universities can support sustainable development by placing emphasis on the development of basic science and technical skills among their graduates. Key to the practice of corporate environment, health, and safety and to the practice of sustainable development is a thorough working knowledge of how various systems, both ecological and institutional, work and interact. Without this broad general framework, individual actors cannot be expected to make decisions that are consistent with societal goals.

Universities can place high value on working as part of an interdisciplinary team. Particularly at the graduate level, students can be given assignments that require them to work as a contributor to a group whose members have differing academic training. These exercises will help prepare students for working in an environment in which the "culture" and "language" are different from those in which they are formally trained and will also help them develop an understanding that it is rare when a single discipline has all the answers to a complex problem.

Engineers dealing with technical environmental issues can be taught to communicate those issues to business professionals in a clear and succinct manner. Similarly, both engineers and business professionals can be assisted in recognizing and integrating important issues from each other's fields into their own work.

Faculty and staff at universities can be encouraged to engage in exercises comparable to those of students. Team teaching of courses on contemporary and complex issues can be rewarded, and rules for promotion and tenure can be modified to place value on the contributions of both the individual *and* the individual as part of a team.

CONCLUSIONS

During this era of increasing public concern over environmental issues, U.S. multinationals are developing strategies for responding to new competitive pressures as many important markets change and globalize. Corporate responses to the larger business forces could, in theory, maintain competitiveness and at the same time include advances in the way EHS issues are addressed. Sustainable development could become part of a global corporation's culture, and regionally and locally appropriate decisions in support of the environment could be made by relatively autonomous units of large corporations. Although this outcome is a managerial and technical possibility and a few seeds have been planted, it does not represent the current reality in the corporations we studied.

Significant reorganization and downsizing are creating the need for new overall management strategies. EHS policy and management structure are inextricably interwoven in this process. However, our study identified few actions in support of sustainable development that were not required by regulations, needed within the company to achieve or ensure regulatory compliance, anticipated as becoming the subject of regulation or increased public concern, or requested by host country governments as part of an agreement to do business. When voluntary actions supporting sustainable development were identified, the majority were taken on an ad hoc basis at the initiative of a local manager.

Government requirements and liability concerns were identified as crucial motivators for environmental action in the United States and, increasingly, outside the country. From these observations, the conclusion can be drawn that government should examine ways to strengthen and rationalize its regulatory systems so that the regulated community will more consistently frame decisions so that the goals of sustainable development—which include economic growth—are supported.

There is a lag time built into the present situation that can make both companies and the environment vulnerable. Companies have not yet put EHS management systems firmly in place, nor are host country regulatory regimes fully developed or enforced. The study corporations differ in their approaches to this challenge. Some have chosen visible positions as leaders in the field, and others have chosen to follow. Despite these differences in corporations' strategy, management in individual facilities of corporations continues to have an extremely important role in determining how much environmental protection is delivered.

Although the environmental management function has grown in stature and importance in the companies we studied, and environmental reviews have been added at points in the process to help inform business decisions, the necessary integration at the top level of management remains largely unfinished. In the strategic decisions involving resources and markets, environmental issues continue to be secondary. Moreover, a real tension exists between the need to protect against liability (which implies control and accountability) and the forces driving management toward increasingly greater autonomy of divisions and facilities. These issues remain to be faced and solved before corporations will be partners in achieving the societal goal of sustainable development.

Encouraging steps are being taken. For example, companies such as General Electric and Digital are undertaking efforts to integrate environmental issues into their overall management and training programs, and several companies are in the process of examining what type of educational background and experience is most effective for people in environmental management positions.

Specific findings related to our study propositions follow.

Type of Business

EHS practices vary from one business group to another within large corporations. A factor of interest at the outset of this study was the type of business in which the corporation was engaged. It was assumed that similarities in environmental practice among companies engaged in natural resource use could be identified, and that these practices would be different from practices among companies engaged in manufacturing. We found that among the four large case companies, characterizations of the corporation as a whole were not useful in explaining EHS practice. For example, cultural differences and attitudes toward environment, health, and safety in a division responsible for logging versus a division responsible for producing paper are vast. Even within manufacturing, people in the corporations we interviewed attributed significantly different attitudes toward environment, health, and safety to divisions producing different products.

On the evidence, then, differences in industrial sector alone cannot reliably predict important differences in EHS program effectiveness. In other words, similarities in EHS program effectiveness need to be explained in terms other than industrial sector.

Name Recognition

Corporations are concerned about whether they are perceived by the public as being environmentally responsible. This study tested the proposition that corporations with high consumer name recognition develop and implement proactive environmental programs to protect their good names. One case company's name is a household word and another is virtually unknown to consumers because the company sells no products to the consumer under the corporate name. Even the corporation with negligible public recognition took actions it said were designed to protect its name, and we believe that no significant differences among programs were related to varying degrees of consumer name recognition among the corporations we studied.

Individuals within companies who believe there are strong business reasons to be protective of the environment will be more motivated to develop good programs than those from companies who do not believe this.

Company Size

Small companies are not necessarily poorer environmental performers than large ones. One of the propositions tested was the impact of company size on environmental practice. The single small company in the study had a less formalized program and had developed fewer of the program components (such as written policies and regular audit programs) than the four larger case companies. However, the small company made voluntary long-term investments in environmental protection, was actively pursuing process changes to minimize waste production, and earned a positive environmental reputation among government regulators and corporate peers. This positive record was not matched in all of the larger case companies.

Profitability

An expectation at the outset of the study was that profitable corporations would have good environmental programs and less profitable corporations would have poorer environmental programs. This proposition was not supported by case company practice. Among the facilities of case companies, a less effective program was identified within a profitable corporation, and a good program was identified within a less profitable corporation. With the increasing decentralization in many corporations, case company results suggest that profitability of the individual facility may be a more reliable indicator of performance than profitability of the corporation as a whole.

Distance from Headquarters

At the outset of this research, it was assumed that facilities located close to corporate headquarters would have the best EHS programs and that program quality would diminish as distance increased, particularly if cultural, economic, and social conditions in host countries were dramatically different from those in the home country. The assumption was found to be invalid among case companies, and facility age was identified as likely to be a better indicator of performance than distance from headquarters.

For example, one case company had a facility within a few miles of its U.S. corporate headquarters that a staff person characterized as being below-average for the corporation, and another case company had an operation in Brazil with EHS programs in place that were absent in the United States. The below-average U.S. facility is 100 years old, the company's oldest, and the Brazilian facility is newer than its U.S. counterpart. Case companies said that when they build new facilities, regardless of location, they design in a variety of features that result in (1) minimizing waste production and worker exposure to toxic compounds, and (2) maximizing ability to respond to spills and toxic releases. The relationship between facility age and EHS performance is an area that warrants further study.

Lack of conformity with the EHS program developed by corporate headquarters need not have a negative outcome for health or environment, nor need it increase corporate liability if an effective management system is in place that emphasizes quality and continuous improvement.

Top Management Commitment

Top management commitment was cited as important to developing and implementing effective EHS programs; however, there was considerable variation across companies in the examples offered as evidence of commitment. Individuals we interviewed explained the differences as being linked to the individual culture of the corporation or the business unit; what works for one corporation, they say, will not necessarily be adequate for another.

The study finds that the correlation between top management commitment and EHS effectiveness is weakened in a decentralized situation unless division managers are also committed.

Well-Publicized Incident

Being associated with a highly publicized environmental incident did not result in the development of aggressive EHS programs in all cases. We expected to find that an important influence in the development of a comprehensive and effective program in corporations was the occurrence of a negative event; however, two of the five case companies had publicized negative events, and the proposition was supported in one company and not in the other.

One case company was associated with two dramatic events and has used them as a rationale for implementing programs with strong central reporting and regular top management attention. Another case company was named in a widely publicized environmental event, but interviews clearly indicate that in at least one division, a short-term "accounting" mentality still dominates top management thinking and that resources for progressive environmental programs are not readily available to facilities.

RECOMMENDATIONS

Corporations wishing to strengthen their EHS performance should first ensure that an effective management system is in place at all levels. By effective we mean emphasizing quality and continuous improvement.

MNCs face significant challenges in the management of environment, health, and safety in their operations around the world. The emphasis on environment, health, and safety in many corporations has been on the technical aspects of these functions; however, the findings of this study suggest that creative and effective *managerial* approaches lie behind good environmental programs.

Establishing management systems that emphasize quality and continuous improvement will increase the likelihood of good environmental performance, particularly in decentralized operations. The challenges associated with operating in diverse cultural, political, and regulatory environments are being addressed in some multinationals by decreasing emphasis on adoption of specific procedures and increasing emphasis on management systems and facility responsibility. Such an approach provides an opportunity for local innovation and adaptation but need not have a negative outcome for health and environment, nor need it increase corporate liability if effective management systems are in place.

Next, EHS considerations should be integrated into the decision-making process at all levels. Environment, health, and safety should be factors in determining not only how products are made but *whether* they are made. Corporations should not wait for governments to tell them that specific products or resource uses are unacceptable; they should develop and implement criteria for assessing new products and business ventures that take into account all aspects of health, safety, and resource use.

In addition to integration, champions for the environment are critical. A highly visible individual with resources to implement effective programs was identified in the case studies as an important factor in managing EHS on a global scale. Our survey data supports this assertion, with 94 percent of the respondents indicating they agreed/strongly agreed with this statement: The existence within the company of an individual with commitment to the issues, influence, and communication skills is essential to an effective EHS program. See Figure 10.2.

The CEO is an environment champion in one of the case companies and in Du Pont. In other companies, the top person identified with EHS is a vice president. Despite the fact that many respondents felt the existence of a high-level person was critical, in response to the survey question asking whether the company actually had a vice president with exclusive responsibility for environment, health, and safety issues, 68 percent responded negatively and 28 percent indicated there was an individual in this position. See Figure 10.3. As indicated earlier, the survey showed that some differences in company practice are linked to the existence of a vice president of environment, health, and safety.

Our experience in observing companies suggests that both top level and local champions are needed if a company is to effectively capture opportunities to excel in environment, health, and safety management.

Additional research should be undertaken leading to the development of new or improved tools for EHS management in corporations. For example, measures of performance that permit comparisons across corporations and within a corporation across diverse businesses will facilitate communication with top management and will help both researchers and practitioners better understand which EHS actions are placebos and which result in advancement.

Finally, both governments and universities must establish conditions that reinforce creative interdisciplinary approaches to improvements in environment, health, and safety. Just as individuals have been assisted in understanding the personal responsibility they have for the environment so that they increasingly

Figure 10.2
The Existence within the Company of an Individual with Commitment Is Important to Effective EHS Practice

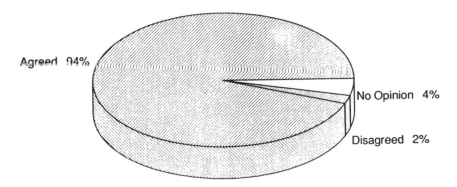

Agreed 94%

No Opinion 4%

Disagreed 2%

Figure 10.3
Do You Have a Vice President with Exclusive Responsibility for Environment, Health, and Safety?

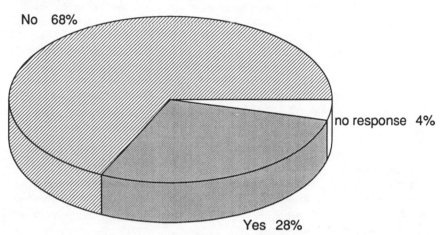

No   68%

no response   4%

Yes   28%

practice recycling, corporations should be encouraged to understand their responsibility for sustainable development. This responsibility includes not only developing and implementing protective EHS programs within the corporation, but also identifying mechanisms for sharing systems and skills with host country governments and with small and medium-size companies in host countries.

Governments should increasingly use a technique that has worked well in industry as a motivator: positive reinforcement. In the field of quality, the Malcom Baldridge award has become a sought-after indicator of progressive corporate management. We believe there are countless examples of positive EHS actions taken by industry that are worthy of government recognition. At the state level, for example, Wisconsin has a governor's award that recognizes corporate pollution prevention efforts and also requires that the approach be written up and shared widely. As a consequence, this program rewards positive actions and simultaneously creates a system for sharing information with other companies.

We believe that in most corporations there are individuals working to change the organizational culture to more fully embrace environment, health, and safety considerations in all aspects of the business. The most effective programs of governments and universities will be designed to recognize and reinforce the actions of these agents for change.

NOTES

1. World Commission on Environment and Development, *Our Common Future* (New York: Oxford University Press, 1987), pp. 8, 42.
2. WCED, *Our Common Future*, p. 65.
3. Ibid., pp. 219–222.

4. Ibid., pp. 222, 223.

5. Ibid., p. 223.

6. United Nations Economic and Social Council, Commission on Transnational Corporations (ECOSOC), *Transnational Corporations and Issues Relating to the Environment*, E/C.10/1990/10, 11 (New York: UN, January 1990), p. 33.

7. ECOSOC, *Transnational Corporations*, pp. 34–41.

8. Stefan H. Robock and Kenneth Simmonds, *International Business and Multinational Enterprises* (Homewood, Ill.: Irwin, 1983), p. 422.

9. ECOSOC, *Transnational Corporations* pp. 32–43.

10. Kenuchi Ohmae, "Planting for a Global Harvest," *Harvard Business Review* (67) 4 (1989): 139.

11. See Michael Dertouzos et al., *Made in America: Regaining the Productive Edge* (Cambridge, Mass.: MIT Press, 1989).

12. Sheldon Novick et al., eds., *Law of Environmental Protection* (New York: Clark Boardman, 1988), pp. 3–9.

13. ICF Incorporated, *Pollution Control Technology Research and Development: Private Sector Incentives and the Federal Role in the Current Regulatory System*, prepared for the U.S. Environmental Protection Agency under contract no. 68-01-6614 (Washington, D.C.: ICF, September, 1984).

14. Federal Technology Transfer Act of 1986, P.L. 99–502, Oct. 20, 1986, 100 Stat. 1785.

15. Executive Order No. 12591, "Facilitating Access to Science and Technology," *Federal Register* (52) (April 10, 1987), p. 13414.

16. Anthony D. Cortese, "Tufts Environmental Literacy Institute Executive Summary," (Medford, Mass.: Office of Environmental Programs, Tufts University, Fall, 1990).

17. See a set of business school cases prepared for the National Wildlife Federation by Forest Reinhardt of Harvard University "Du Pont Freon Products Division (A)" and "Du Pont Freon Products Division (B)," (Washington, D.C.: National Wildlife Federation, 1989).

18. See James E. Post, "The Greening of Management," *Issues in Science and Technology*, Summer 1990, pp. 68–72.

19. An example of failure to consider disposal early in the process is reflected in the recent introduction of the "squeezable" ketchup bottle and the more recent commitment to change the bottle's formulation: "Responding to consumer pressure for more environmentally friendly packaging, the H. J. Heinz Company said yesterday that it would replace its plastic ketchup bottle with another type of plastic that it said was much more likely to be recycled." See John Holusha, "New Plastic in Heinz Bottles to Make Recycling Easier," *New York Times*, April 10, 1990.

Select Bibliography

Ackerman, Robert W. "How Companies Respond to Social Demands." *Harvard Business Review* 51 (July–August 1973): 88–89.

"Allied-Signal's Plant Surveillance Program." *Environmental Manager* 1, no. 9 (1990): 1.

Arthur D. Little, Inc. *Environmental, Health and Safety Policies: Current Practices and Future Trends*. Cambridge, Mass.: Arthur D. Little, Inc., 1988.

Attalah, S. "Assessing and Managing Industrial Risk." *Chemical Engineering* (September 8, 1980): 99–103.

Badaracco, Joseph L. *Allied Chemical Corporation: Harvard Business School Case Study No. 379–137*. Boston: Harvard University, 1979.

Baram, Michael S. *Chemical Industry Accidents: Risk Communication and Emergency Planning in the U.S. and the E. C.* Unpublished manuscript submitted for presentation at the World Conference on Chemical Accidents, Rome, Italy: 7–10 July 1987.

—— *Corporate Risk Management: Industrial Responsibility for Risk Communication in the European Community and the United States*. Brussels: Commission of the European Communities, October 1987.

—— "The Right to Know and the Duty to Disclose Hazard Information." *American Journal of Public Health* 74, no. 4 (April 1984): 380.

—— "Risk Communication Law and Implementation Issues in the United States and the European Community." in *Corporate Disclosure of Environmental Risks: U.S. and European Law*, eds. Michael S. Baram and Daniel G. Partan. Salem, N. H.: Butterworth Legal Publishers, 1990.

Barros, James, and Douglas Johnston. *The International Law of Pollution*. New York: Free Press, 1974.

Beecher, Norman and Ann Rappaport. "Hazardous Waste Management Policies Overseas." *Chemical Engineering Progress* (May 1990): 30–39.

Behrman, Jack N. *Conflicting Constraints on the Multinational Enterprise: Potential for Resolution*. New York: N.Y. Council of the Americas, 1974.

—— *Industry Ties with Science and Technology Policies in Developing Countries*. Cambridge, Mass: Oelg Eschlager, Gunn and Hain, 1980.

—— *U.S. International Business and Governments*. New York: McGraw-Hill, 1971.

Behrman, Jack N., and William Gilbert Carter. *Problems of International Business Cooperation in Environmental Protection*. New York: Fund for Multinational Management Education, 1975.

Bendiner, Burton. *International Labor Affairs: The World Trade Unions and the Multinational Companies*. New York: Oxford University Press, 1987.

Benedick, Richard E. *Environment in the Foreign Policy Agenda*. Washington, D.C.: U.S. Department of State, Bureau of Public Affairs, 1986.

Bergsten, C. Fred; Thomas Horst, and Theodore H. Moran. *American Multinationals and American Interests*. Washington, D.C.: The Brookings Institute, 1978.

Bothe, Borut. "UNEP's Environmental Law Activity on International Transport and Disposal of Toxic and Dangerous Wastes." *Industry and Environment* 6, no. 4 (1983): 3–6, 18.

Bowonder, B., J. Kasperson and R. Kasperson. "Avoiding Future Bhopals." *Environment* 27, no. 7 (September 1985): 6–37.

Brooke, Michael Z., and Lee H. Remmers. *The Strategy of Multinational Enterprise: Organization and Finance*. New York: American Elsevier, 1970.

Brunner, David L., Will Miller and Nan Stockholm. *Corporations and the Environment: How Should Decisions Be Made?* Los Altos, Calif.: Committee on Corporate Responsibility, Graduate School of Business, Stanford University, 1980.

Business International Corporation. *161 More Checklists: Decision Making in International Operations*. New York: Business International Corporation, 1985.

Buttel, Frederick H.; Charles C. Geisler, and Irving W. Wiswall, eds. *Labor and Environment*. Westport, Conn: Greenwood Press, 1984.

Caldwell, Lynton. *In Defense of the Earth: International Protection of the Biosphere*. Bloomington, Ind.: University Press, 1972.

—— *International Environmental Policy: Emergence and Dimensions*. Durham, N.C.: Duke University Press, 1984.

Carpenter, George. "First Volume." *GEMI-News* (September 1990): 1.

Castleman, Barry I. "Export of Hazardous Factories to Developing Nations." *International Journal of Health Services* 9 (1979).

—— *Workplace Health Protection Standards and Multinational Corporation Activities in Developing Countries*. Paper Prepared for World Resources Institute Conference on the Role of Multinational Corporations in Developing Countries. Washington, D.C.: 15–16 June 1984.

Chemical Manufacturers Association. *International Health and Safety Affairs* (Periodical, Washington, D.C.).

—— *Responsible Care, A Public Commitment: Questions and Answers about Responsible Care*. Washington, D.C.: Chemical Manufacturers Association, April 1991.

Chlorine Institute. *Chlorine: A Guide for Journalists, Pamphlet 70*. Washington, D.C.: Chlorine Institute, February 1980.

Cohen, A. Smith, Michael J. Colligan and Philip Berger. "Psychology in Health Risk Messages for Workers." *Journal of Occupational Medicine* 27 (August 1985).

Cohen, Alexander. "Factors in Successful Occupational Safety Programs." *Journal of Safety Research* 9 (December 1977).

Cohen, Michael J. "Handling Your Multinational Insurance Needs with Bhopal in Mind." *Business Insurance* 17 (1984).

Cortese, Anthony D. *Tufts Environmental Literacy Institute Executive Summary.* Medford, Mass.: Office of Environmental Programs, Tufts University, Fall 1990.

Crean, John G. *The International Community and the Multinational Enterprise: Response and Regulation.* Toronto: Canadian Institute of International Affairs, 1982.

De Bodinat, H. *Influence in the Multinational Corporation: The Case of Manufacturing* D.B.A. diss., Graduate School of Business Administration, Harvard University, 1975.

Deloitte & Touche and Stanford University Graduate School of Business, Public Management Program. *The Environmental Transformation of U. S. Industry: A Survey of U.S. Industrial Corporations; Environmental Strategies, Management Policies and Perceptions*, 1990.

Deming, W. Edwards. *Out of the Crisis.* Cambridge, Mass.: MIT Center for Advanced Engineering Study, 1986.

Dertouzos, Michael L. et al. *Made in America: Regaining the Productive Edge.* Cambridge, Mass.: MIT Press, 1989.

Dillon, Patricia S., and Kurt Fischer. *Environmental Management in Corporations: Methods and Motivations.* Medford, Mass.: Center for Environmental Management, Tufts University, 1991.

Directorate of Intelligence. *Handbook of Economic Statistics 1990.* Washington, D.C.: CIA, 1990.

Duerksen, Christopher J. *Environmental Regulation of Industrial Plant Siting.* Washington, D.C.: The Conservation Foundation, 1983.

Duerr, Michael. *Organization and Control of International Operations.* New York: The Conference Board, 1973.

Dun's Marketing Service. "American Corporate Families." Parsippany, N.J.: Dun and Bradstreet, Co., 1989.

Durkee, Linda C. "Risk Communication and the Rhine River." *International Environment Reporter* 12 (11 October 1989): part II.

Economic and Social Commission for Asia and the Pacific. *Transnational Corporations and Environmental Management in Selected Asian and Pacific Developing Countries, ESCAP/UNCTC Publication Series B, No. 13.* Bangkok: ESCAP/UNCTC, 1988.

El-Hinnawi, Essam, and Manzur H. Hashmi. *The State of the Environment.* Boston, Mass.: United Nations Environment Program and Butterworth, 1987.

"Environmental Requirements of the World Bank." *The Environmental Professional* 7 (1985): 205–212.

Evan, Harry Z., ed *Employers and the Environmental Challenge.* Geneva: ILO/UNEP, 1986.

Fayerweather, John. *Host National Attitudes toward Multinational Corporations.* New York: Praeger, 1974.

Fisher, Bart S., and Jeff Turner. *Regulating the Multinational Enterprise: National and International Challenges.* New York: Praeger, 1984.

Flaherty, Margaret, and Ann Rappaport. *Multinational Corporations and the Environment: A Survey of Global Practices.* Medford, Mass.: Center for Environmental Management, Tufts University, 1991.

Forster, Malcom J. "Hazardous Waste—Toward International Agreement." *Environmental Policy and Law* 12, no. 3 (1984): 64–67.

Friedman, Frank B. *Practical Guide to Environmental Management.* Washington, D.C.: Environmental Law Institute, 1988.

General Electric Corporation. "A Tool to Increase Competitive Advantage: Environmental, Health and Safety Regulations." *EHS News,* (Fall 1989).

Ghertman, Michel. *Decision-Making in Multinational Enterprises: Concepts and Research Approaches,* ILO Working Paper no. 31. Geneva: ILO, 1984.

Gladwin, Thomas N. *Environment, Planning and the Multinational Corporation.* Greenwich, Conn.: JAI Press, 1977.

—— "Environmental Policy Trends Facing Multinationals." *California Management Review* 20 (Winter 1977): 81–93.

Gladwin, Thomas N., and Michael G. Royston. "An Environmentally Oriented Mode of Industrial Project Planning." *Environmental Conservation* 2 (Autumn 1975): 189–198.

Gladwin, Thomas N., and Ingo Walter. "Multinational Enterprise, Social Responsiveness, and Pollution Control." *Journal of International Business Studies* (Fall-Winter 1976): 57–72.

Gladwin, Thomas N., and John G. Welles. "Environmental Policy and Multinational Corporate Strategy." In *Studies in International Environmental Economics,* ed. Ingo Walter. New York: John Wiley, 1976.

Gladwin, Thomas N., and John G. Welles. "Multinational Corporations and Environmental Protection: Patterns of Organizational Adaptation." *International Studies of Management and Organization* 6, no. 1 (1976): 160–184.

Gleckman, Harris. "Proposed Requirements for Transnational Corporations to Disclose Information on Product and Process Hazards." In *Corporate Disclosure of Environmental Risks: U.S. and European Law,* eds. Baram, Michael S. and Daniel G. Partan. Salem, N. H.: Butterworth Legal Publishers, 1990.

Gray, S. J.; C. B. McSweeney, and J. C. Shaw. *Information Disclosure and the Multinational Corporation.* New York: Wiley, 1984.

Griffiths, Richard F., ed. *Dealing with Risk: The Planning, Management and Acceptability of Technological Risk.* New York: Wiley, 1981.

Guile, Bruce R., and Harvey Brooks. *Technology and Global Industry: Companies and Nations in the World Economy.* Washington, D.C.: National Academy Press, 1987.

Haigh, N. *Co-ordination of Standards for Chemicals in the Environment.* Bonn, Federal Republic of Germany: Institute for European Environmental Policy, 1986.

Halter, Faith. "Regulating Information Exchange and International Trade in Pesticides and Other Toxic Substances to Meet the Needs of Developing Countries." *Columbia Journal of International Law* 12, no. 1 (1987).

Harris, Philip R. *Management in Transition: Transforming Managerial Practices and Organizational Strategies for a New Work Culture.* San Francisco: Jossey-Bass, 1985.

Harrison, L. Lee, ed. *The McGraw-Hill Environmental Auditing Handbook.* New York: McGraw-Hill Book Co., 1984.

Hinrichsen, Don. "Like First World, Like Third World." *The Amicus Journal* 10 (Winter 1988).

Holusha, John. "New Plastic in Heinz Bottles to Make Recycling Easier." *New York Times,* (10 April 1990).

ICF Incorporated. "Pollution Control Technology Research and Development: Private Sector Incentives and the Federal Role in the Current Regulatory System." ICF, Washington, D.C.: September 1984.

ILO. *Employment and Technological Choice of Multinational Enterprises in Developing Countries, Working Paper no. 23.* Geneva: ILO, 1983.

—— *Encyclopedia of Occupational Health and Safety,* 2 vols. Geneva: ILO, 1983.

—— *Multinationals' Training Practices and Development.* Geneva: ILO, 1981.

—— *Safety and Health Practices of Multinational Enterprises.* Geneva: ILO, 1984.

—— *Safety, Health and Working Conditions in the Transfer of Technology to Developing Countries, and ILO Code of Practice.* Geneva: ILO, 1988.

International Chamber of Commerce. *The Business Charter for Sustainable Development, Principles for Environmental Management, Pub. No. 210/356A.* Paris: International Chamber of Commerce, April 1991.

—— *Environmental Guidelines for World Industry.* Paris: International Chamber of Commerce, 1990.

Ives, Jane H. *International Occupational Safety and Health Resource Catalogue.* New York: Praeger, 1981.

"Joint U. S., Mexican Manufacturing Program May Be Causing Pollution in Texas, Arizona." *International Environment Reporter,* (June 1988): 306.

Kasperson, Roger et al. *Corporate Management of Health and Safety Hazards: A Comparison of Current Practices.* Boulder: Westview Press, 1988.

Klein, Harold E. and Robert E. Linneman. "Environmental Assessment: An International Study of Corporate Practice." *Journal of Business Strategy* 5 (Summer 1984): 66–75.

Kline, John M. *International Codes and the Multinational Business: Setting Guidelines for International Business Operations.* Westport, Conn.: Quorum Books, 1985.

Knodgon, Gabriele. *Environment and Industrial Siting: Results of an Empirical Survey of Investment by West German Industry in Developing Countries.* Berlin: International Institute for Environment and Society, May 1979.

Koza, Mitchell et al. "Company Policies for Environmental Protection: A Preliminary Study of Nine European Companies." December 1989. Report prepared for presentation at a meeting on Public Information: Companies' Organization to Deal with Environmental Issues. Paris: UNEP, Industry and Environment Office, December 1989.

LaDou, Joseph. "Deadly Migration." *Technology Review* (July 1991): 49.

Lafferre, T. H. Speech to American Institute of Chemical Engineers, Conference on Waste Minimization, Washington, D.C., 4 December 1989.

Lall, S. "Transnationals, Domestic Enterprises, and Industrial Structure in Host LDCs: A Survey." *Oxford Economic Papers* 30 (July 1978).

Leonard, H. Jeffrey. *Are Environmental Regulations Driving U.S. Industry Overseas?* Washington D.C.: Conservation Foundation, 1984.

—— "Confronting Industrial Pollution in Rapidly Developing Countries: Myths, Pitfalls, and Opportunities." *Ecology Law Ouarterly* (1985): 779–783.

—— *Divesting Nature's Capital: The Political Economy of Environmental Abuse in the Third World.* New York: Holmes and Mayer, 1985.

—— "Environmental Regulations, Multinational Corporations and Industrial Development in the 1980s." *Habitat International* 6, no. 3 (1982):

—— "Pollution Plagues Industrial Firms in Growing Nations." *Conservation Foundation Letter* (August 1982).

Leonard, H. Jeffrey and Christopher J. Duerksen. "Environmental Regulations and the Location of Industry: An International Perspective." *Columbia Journal of World Business.* (Summer 1980).

Leonard, H. Jeffrey and David Morell. "Emergence of Environmental Concern in Developing Countries: A Political Perspective." *Stanford Journal of International Law* 17, no. 2 (1981): 281–313.

Leonard, Richard. "After Bhopal: Multinationals and the Management of Hazardous Products and Processes." *Multinational Business* 2 (1986).

Levenstein, Charles and Stanley Eller. "Are Hazardous Industries Fleeing Abroad?" *Business and Society Review* (Summer 1980).

Lund, Leonard. *Corporate Organization for Environmental Policymaking.* New York: The Conference Board, 1974.

Lydenberg, Steven D., Alice Tepper Marlin, Sean O'Brien Strub, and the Council on Economic Priorities. *Rating America's Corporate Conscience: A Provocative Guide to the Companies Behind the Products You Buy Every Day.* Reading: Addison-Wesley, 1986.

MacDonald, Donald L. *Risk Control in the Overseas Operations of American Corporations.* Ann Arbor: University of Michigan, 1979.

Madu, Christian N. "Transferring Technology to Developing Countries—Critical Factors for Success." *Long Range Planning* 22 (August 1989): 115–124.

Modelski, George. *Transnational Corporations and World Order: Readings in International Political Economy.* San Francisco: W. H. Freeman and Company, 1974.

Moody's Industrial Manual. New York: Moody's Investors Service, 1988.

Nadis, Steve. "Mexican Clean-up." *Technology Review* (November/December 1989): 10.

Nagel, Stuart S. "Incentives for Compliance with Environmental Law." *American Behavioral Science* 17 (May–June 1974).

Natkin, Alvin M. "Once Is Too Often: Corporate Responsibility in the Aftermath of Bhopal." *Journal '85* World Resources Institute (1985): 62–67.

Negandhi, A. R. *Organization Theory in an Open System: A Study of Transferring Advanced Management Practices to Developing Nations.* New York: Dunellen, 1975.

Nelkin, D., and M. S. Brown. *Worker at Risk: Voices from the Workplace.* Chicago: University of Chicago Press, 1984.

Novick, Sheldon, et al., eds. *Law of Environmental Protection.* New York: Clark Boardman, 1988.

O'Neill, Larry. *One Year Later: Report of the Monsanto Product and Plant Safety Task Force.* St. Louis: Monsanto, 1985.

OECD, "Clarification of Environmental Concerns in OECD Guidelines for Multinational Enterprises." in *International Legal Materials* 25 (1986): 494.

—— "Decision and Recommendation on Transfrontier Movements of Hazardous Waste." In *International Legal Materials* 23 (1 February 1984).

—— "Documents from the High Level Meeting on Chemicals." In *International Legal Materials* 19, no. 4 (1980).

—— *Environmental Policy and Technical Change.* Paris: OECD, 1985.

—— "OECD Guidelines on Multinational Enterprises." In *International Legal Materials* 15 (1976): 969–976.

—— *Recommendations of the Council on Guiding Principles Concerning International Economic Aspects of Environmental Policies*. Paris: OECD, 1977.

—— *Responsibility of Parent Companies for their Subsidiaries*. Paris: OECD, 1980.

Ohmae, Kenichi. "Planting for a Global Harvest." *Harvard Business Review* 67, no. 4 (1989): 136–145.

Paden, Mary, ed. *World Resources and U.S. Interests: Business' Stake*. Washington, D.C.: World Resources Institute, 1985.

Parker, J. E. S. *The Economics of Innovation: The National and Multinational Enterprise in Technological Change*. London: Longman, 1974.

Partan, Daniel G. "The Duty to Inform in International Law." in *Corporate Disclosure of Environmental Risks: U.S. and European Law*, eds. Baram, Michael S. and Daniel G. Partan. Salem, N.H.: Butterworth Legal Publishers, 1990.

Pearson, Charles S. *Down to Business: Multinational Corporations, the Environment, and Development*. Washington, D.C.: World Resources Institute, January 1985.

—— *Multinational Corporations, Environment, and the Third World: Business Matters*. Durham: Duke University Press, 1987.

—— "What Has to Be Done to Prevent More Bhopals." *Journal '85* World Resources Institute (1985): 58–61.

Pearson, Charles, et al. *Improving Environmental Cooperation: The Roles of Multinational Corporations and Developing Countries*. The Report of a Panel of Business Leaders and Other Experts Convened by the World Resources Institute. Washington, D.C.: World Resources Institute, 1985.

Pease, Elizabeth Sue. *Occupational Safety and Health: A Sourcebook*. New York: Garland Publishers, 1985.

Penna, Frank J. *A New Alliance: Cooperation between PVOs and Multinational Corporations-Report of a Conference*. New York: Fund for Multinational Management Education, 7 October 1986.

Petillot, F. "The Policies and Methods Established for Promoting the Development of Clean Technologies in French Industry." *Industry and Environment* 4 (1986).

Phatak, Arving V. *Managing Multinational Corporations*. New York: Praeger, 1974.

Pollio, Gerald and Charles Riemenschneider. "The Coming Third World Investment Revival." *Harvard Business Review* (March–April 1988).

Post, James E. "The Greening of Management." *Issues in Science and Technology* (Summer 1990): 68–72.

Poynter, Thomas. *Multinational Enterprises and Government Intervention*. New York: St. Martin's Press, 1985.

Prabhu, Mohan A. *Protection against Chemical Hazards: A Comparative Examination of the Laws of the United Kingdom, France, the Federal Republic of Germany, the European Economic Community, the United States of America, and Japan, and Related Conventions and Activities of International Organizations*. Ottawa, Canada: Department of Justice, 1988.

Rand McNally. *World Facts in Brief*. Chicago, Ill.: Rand McNally & Company, 1986.

Reid, Walter V.; James N. Barnes, and Brent Blackwelder. *Bankrolling Successes: A Portfolio of Sustainable Development Projects*. Washington, D.C.: Environmental Policy Institute and National Wildlife Federation, 1988.

Reinhardt, Forest. *Du Pont Freon Products Division*. Washington, D.C.: National Wildlife Federation, 1989.

Remick, David. "Soviets Report 250 Deaths Occurred at Chernobyl; Official Toll Following 1986 Nuclear Accident in the Ukraine Had Been Only 31." *Washington Post,* (9 November 1989): A70.

Research and Development Scoreboard. "A Perilous Cutback in Research Spending." *Business Week* (20 June 1988).

Robock, Stefan H., and Kenneth Simmonds. *International Business and Multinational Enterprises.* Homewood, Ill.: Irwin, 1983.

Rohter, Larry. "The Makesicko City, New Smog Fear." *New York Times,* (12 April 1989): 4.

Rosencranz, A. "Bhopal, Transnational Corporations and Hazardous Technologies." *Ambio* 17, no. 5 (1988): 336–340.

Rostow, Eugene V. and George W. Ball. "The Genesis of the Multinational Corporation." In *Global Companies: The Political Economy of World Business,* ed. Ball, George W. Englewood Cliffs, NJ: Prentice Hall, 1975.

Royston, Michael G. "Control by Multinational Corporations: The Environmental Case for Scenario 4." *Ambio* 7, no. 2/3 (1979).

Royston, Michael G. *The Role of Multinational Corporations in Environment and Resource Management in Developing Countries-Paper Presented at the World Resources Institute Conference on the Role of Multinational Companies in Environmental Management in Developing Countries.* Washington, D.C.: 14–16 July 1984.

Sallada, Logan H., and Brendan G. Doyle. *The Spirit of Versailles: The Business of Environmental Management.* Paris: ICC Publishing SA, 1986.

Shaikh, Rashid A. "The Dilemmas of Advanced Technology for the Third World." *Technology Review* 89, no. 3 (1986): 57.

Shaw, Patricia M. "International Legislation and the Transport of Hazardous Wastes." *Industry and the Environment* (UNEP) 4 (1983): 63–65.

Sheehan, Arlene M. "Chemical Plant Safety Requirements: The European Example." *Law and Policy in International Business* 16, no. 2 (1984): 621–640.

Solomon, Jolie. "U.S. Firms' Standards in Mexico Targeted." *The Boston Globe* (13 February 1991): 29–30.

Springer, Allen. *The International Law of Pollution: Protecting the Global Environment in a World of Sovereign States.* Westport, Conn.: Quorum Books, 1983.

Staaf, Robert J., and Francis X. Tannian. *Externalities: Theoretical Dimensions of Political Economy.* New York: Dunellen, 1973.

Sundram, S. T. "Malaysia's Approach to Environmental Management." *Industry and Environment* (August–September 1984).

Szekely, Francisco. "Pollution for Export." *Mazingira* 3/4 (1977).

Taubenfeld, Howard J., and Virginia E. Templeton. *World Environmental Law Bibliography.* Littleton, Colo.: Fred B. Rothman & Co., 1987.

Taylor, J. Gary. *Environmental Planning in the Context of Development Investment.* Medford, Mass.. Department of Urban and Environmental Policy, Tufts University, 1983.

—— *Managing Environmental Risk in Newly Industrializing Countries.* Medford Mass.: Department of Urban and Environmental Policy, Tufts University, 1987.

Tosato, M. *Recent National and International Approaches to the Control of Existing Chemicals: Proceedings of the International Workshop No. 2.* Rome, Italy: Instituto Superiore di Sanita, 1986.

Trisoglio, Alex and Kerry ten Kate. *From WICEM to WICEM II: A Report to Assess Progress in the Implementation of the WICEM Recommendations.* Paris: United Nations Environment Programme, Industry and Environment Office, March 1991.

Tund, Rosalie T., ed. *Strategic Management in the United States and Japan: A Comparative Analysis.* Cambridge, Mass.: Ballinger Publishing Co., 1986.

Uhlig, Mark A. "Mexico Closes Giant Oil Refinery to Ease Pollution in the Capital." *New York Times* (19 March 1991): A1.

—— "Refinery Closing Outrages Mexican Workers." *New York Times* (27 March 1991): A11.

UNEP. "Environmental Data Report, Second Edition, 1989–90." *Oxford, UK: Basil Blackwell Ltd.* (1989).

"Union Carbide Fights for Its Life." *Business Week,* (24 December 1984): 53–56.

United Nations. "Draft Code of Conduct on Transnational Corporations." In *International Legal Materials* 23 (1983): 626–640.

—— "Protocol on Substances That Deplete the Ozone Layer, Montreal, 16 September 1987." In *International Legal Materials* 26 (1987): 1541–1561.

United Nations Center on Transnational Corporations. *Environmental Aspects of the Activities of Transnational Corporations: A Survey ST/CTC55. UNPUB. E.85.II. A.11.* New York: United Nations, 1985.

—— *Further International Co-Operation for Environmental Management of Industrial Process Safety and Hazards: Informal Seminar UNCTC Overview Paper.* Geneva: United Nations, 12 May 1985.

—— *Transnational Corporations and Technology Transfer: Effects and Policy Issues.* New York: United Nations, 1987.

United Nations Economic and Social Council, Commission on Transnational Corporations. *Transnational Corporations and Issues Relating to the Environment (E/C.10/1990/10,* United Nations, New York: January, 1990.

United Nations Environment Programme Caribbean Regional Co-ordinating Unit. *Action Plan for the Caribbean Environment Programme.* Kingston, Jamaica: UNEP, October 1987.

United Nations Environment Programme, Industry and Environment Office. *APELL, Awareness and Preparedness for Emergencies at Local Level U.N. Pub. No. E.88.III.D.3.* Paris: United Nations, 1988.

——. "Technological Accidents No. 1." *Industry and Environment* 11, no. 2 (April/May–June 1988).

——. "Technological Accidents No. 2." *Industry and Environment* 11, no. 3 (July–August/September 1988).

United Nations Environment Programme. *Crisis, External Debt, Macroeconomic Policies and their Relation to the Environment in Latin America and the Caribbean,* UNEP, Brasilia, Brazil: 30–31 March 1989.

—— *Debt-for-Conservation Swaps in Latin America and the Caribbean,* UNEP, Brasilia, Brazil: 27–29 March 1989.

United Nations Environment Program. *Draft Environmental Perspective,* UNEP, 30 April 1987.

—— *Environmental Law Guidelines and Principles Exchange of Information on Chemicals in International Trade,* UNEP, Nairobi, Kenya: 17 June 1987.

United Nations Environment Programme. *Final Report of the High-Level Expert Meeting on Environmental Management in Latin America,* UNEP, Brasilia, Brazil: 27–29 March 1989.

United Nations Environment Programme. *The Global 500: The Roll of Honor for Environmental Achievement.* UNEP, 1989.

——— *Guidelines on Risk Management and Accident Prevention in the Chemical Industry.* Moscow: UNEP, 1982.

——— *Guidelines for Environmental Management of Iron and Steel Works.* Moscow: UNEP, 1986.

——— *Industry and Environment* (quarterly): Paris.

United Nations Environment Programme. *Part I: Assessment of Regional and Global Environmental Trends and Issues: Implications for a Regional Policy,* UNEP, Brasilia, Brazil: 27–29 March 1989.

——— *Project Progress Report on the Development of Environmental Legislation and Institutional Frameworks (PR-5),* UNEP, Brasilia, Brazil: 27–29 March 1989.

——— *Provisional Thematic Index of Environmental Legislation in Force in Latin America and the Caribbean,* UNEP, Brasilia, Brazil: 27–29 March 1989.

——— *Register of International Treaties and other Agreements in the Field of the Environment,* UNEP, Nairobi, Kenya: May 1985.

United Nations Environment Programme. *UNEP Profile.* Nairobi, Kenya: UNEP, 1987.

United Nations Environment Program. "WICEM Outcomes and Reactions." *Industry and Environment: Special Issue* (1984).

U.S. Congress, Office of Technology Assessment. "Serious Reduction of Hazardous Waste: For Pollution Prevention and Industrial Efficiency." Washington, D.C.: U.S. Government Printing Office, 1986.

U.S. Department of State, Bureau of Public Affairs, Office of Public Communication, Editorial Division. *Background Notes, Brazil.* Washington, D.C.: U.S. Department of State, 1987.

U.S. Environmental Protection Agency. *Risk Assessment and Risk Management: Framework for Decision Making.* Washington, D.C.: U.S. Environmental Protection Agency, December 1984.

——— "Title III Fact Sheet, Emergency Planning and Community Right-to-Know." Washington, D.C.: U.S. Environmental Protection Agency, August 1988.

United States of Mexico, Ministry of Urban Development and Ecology. *General Law for Environment Protection and Ecological Equilibrium,* U.S. Mexico, Mexico City: December 1987.

Valentino, F. W., and G. E. Walmet. "Industrial Waste Reduction: The Process Problem." *Environment* 28, no. 7 (1986).

Vandamme, Jacques, ed. *Employee Consultation and Information in MNCs.* London: Dover, 1986.

Vernon, Raymond. *The Economic and Political Consequences of Multinational Enterprise: An Anthology.* Boston: Harvard Business School, 1972.

Vernon, Raymond. "Storm Over the Multinationals: Problems and Prospects." *Foreign Affairs* (January 1977).

Vertinsky, I., and P. Vertinsky. "Communicating Environmental Health Assessment and other Risk Information: Analysis of Strategies." *Risk: a Seminar Series* H. Kunreuther ed., Laxenberg, Austria: International Institute for Applied Systems Analysis (1982).

Viola, Eduardo. "The Ecologist Movement in Brazil, 1974–1986: From Environmentalism to Ecopolitics." *International Journal of Urban and Regional Research* 12 (1988).

Vogel, David. *National Styles of Regulation: Environmental Policy in Great Britain and the United States.* Ithaca, N.Y.: Cornell University Press, 1986.

Von Moltke, Konrad. "Bhopal and Seveso-Avoiding a Recurrence." *The Environmental Forum* (June 1985): 21–23.

Wallace, Don. *International Regulation of Multinational Companies.* New York: Praeger, 1976.

Walter, Ingo. *Environmental Control and Consumer Protection: Emerging Forces in Multinational Corporate Operations.* Washington, D.C.: Center for Multinational Studies, 1972 2d. ed., 1975.

—— *Studies in International Environmental Economics.* New York: Wiley, 1976.

Weiss, Bernard, and Thomas W. Clarkson. "Toxic Chemical Disasters and the Implications of Bhopal for Technology Transfer." *Milbank Quarterly* 61, no. 2 (1986): 216–240.

Welles, John G. "Multinationals Need New Environmental Strategies." *Columbia Journal of World Business* (Summer 1973): 11–18.

Whipple, Chris, and Vincent T. Covello, eds. *Risk Analysis in the Private Sector.* New York: Plenum Press, 1985.

White, Allen. "The Transboundary Movement of Hazardous Products, Processes and Wastes from the US to Third World Nations." Baltimore, M.D., 20 March 1989. Paper Presented at the Annual Meeting of the Association of American Geographers.

"Who's Where in Profitability." *Forbes,* (9 January 1989).

Woolard, E. S. *Corporate Environmentalism, Remarks by the chairman, Du Pont, before the American Chamber of Commerce London, May 4, 1989.*

World Bank. *Draft World Bank and IFC Guidelines for Identifying and Analyzing Major Hazard Installations in Developing Countries.* Washington, D.C.: World Bank OESA, February 1985.

—— *Manual of Industrial Hazard Assessment Techniques.* Washington, D.C.: World Bank OESA, 1985.

World Commission on Environment and Development. *Our Common Future.* New York: Oxford University Press, 1987.

World Health Organization. "Risk Management in Chemical Safety." *Science of the Total Environment* 51 (1986).

World Resources Institute. *Improving Environmental Cooperation: The Roles of Multinational Corporations and Developing Countries,* World Resources Institute, Washington, D.C.: 1984.

Yin, Robert K. *Case Study Research: Design and Methods.* Beverly Hills: Sage Publications, 1984.

"3M Announces Plan to Cut Hazardous Releases by 90 Percent, Emphasize Pollution Prevention." *Environmental Reporter,* (16 June 1989): 441.

Index

References to case company names are *italicized*.

ABOUT THE AUTHORS

ANN RAPPAPORT is Senior Environmental Research Analyst at the Center for Environmental Management, Tufts University. She also has academic appointments in the departments of civil engineering and urban and environmental policy at Tufts. Rappaport is a former government official who helped develop and implement the hazardous waste regulatory program in Massachusetts. She has published articles in *Technology Review, Chemical Engineering Progress,* and *International Environment Reporter.*

MARGARET FRESHER FLAHERTY is Environmental Research Analyst at the Center for Environmental Management, Tufts University. Flaherty has been involved in research in the area of corporate environmental programs and is currently examining the environmental issues of state-owned enterprises in Brazil and Hungary. She has recently published *Multinational Corporations and the Environment: A Survey of Global Practices* as well as an article in *International Environment Reporter.*